Stable Gas-in-Liquid Emulsions

Production in Natural Waters
and Artificial Media

Second Edition

STUDIES IN INTERFACE SCIENCE

SERIES EDITORS
D. Möbius and R. Miller

Stable Gas-in-Liquid Emulsions
Production in Natural Waters
and Artificial Media

Second Edition

Joseph S. D'Arrigo

Cavitation-Control Technology, Inc. (CAV-CON)
55 Knollwood Road, Farmington, CT 06032, U.S.A.

2003
ELSEVIER
Amsterdam – Boston – Heidelberg – London – New York – Oxford
Paris – San Diego – San Francisco – Singapore – Sydney – Tokyo

ELSEVIER SCIENCE B.V.
Sara Burgerhartstraat 25
P.O. Box 211, 1000 AE Amsterdam, The Netherlands

First edition 1986
Second edition , first printing 2003

Library of Congress Cataloging in Publication Data
A catalog record from the Library of Congress has been applied for.

British Library Cataloguing in Publication Data
A catalogue record from the British Library has been applied for.

ISBN: 0 444 51482 1
ISSN: 1383 7303

⊚ The paper used in this publication meets the requirements of ANSI/NISO Z39.48-1992 (Permanence of Paper).
Printed in Hungary.

To
Sachie, Paul, and Marie

PREFACE - Second Edition

      Stable gas-in-liquid emulsions, as found in natural waters or when modeled from natural microbubbles using artificial media, are basically coated microbubbles and represent one more example of "self-assembly" in science.

      As pointed out in a recent Special Feature issue published by the National Academy of Sciences (USA), even though the concepts of "self-assembly" were developed with molecules, the expanding contact of chemistry with biology and materials science as well as the direction of technology toward nanometer- and micrometer-scale structures, however, have widened this focus to include self-assembling processes at scales larger than the molecular. There are now three ranges of sizes of components for which self-assembly is important: molecular, nanoscale (colloids, nanospheres, and related structures), and meso- to macroscopic (objects with dimensions from microns to larger) (G.M. Whitesides & M. Boncheva (2002) Proc. Natl. Acad. Sci. USA 99:4769-4774).

      The surfactant-coated microbubbles described in this book range in size from nanoscale (i.e., submicron) to mesoscale (i.e., microns or micrometers), and fall into two categories. First, surfactant-stabilized natural microbubbles (~ 0.5-100 μm in diameter), also referred to as dilute gas-in-liquid emulsions, are reviewed and analyzed in Chapters 1-8 of the book. Second, the synthetic or artificially coated microbubbles (from submicron to a few micrometers in diameter), also referred to as concentrated gas-in-liquid emulsions or as lipid-coated microbubbles, are described and their properties examined in detail in Chapters 9-15.

      The first ten chapters, which comprised the earlier (1986) edition of the book, have now been updated (with additional, more recent references) in the current expanded edition -- and Chapter 10 has undergone some revision.

      Chapters 11-15 are entirely new; these five chapters review

over one and a half decades of research from, or in collaboration with, this laboratory investigating lipid-coated microbubbles (LCM) and their medical applications. Much experimental data is presented and examined in detail, along with the relevant current literature, which collectively reveal that the particular lipid coating of LCM causes them to display marked tumor-targeting abilities -- useful for both diagnosis and treatment of tumors. Numerous recent literature references related to these added R&D programs have been included and the (subject) index updated.

Thanks are due to the following colleagues for their collaboration on some of the original investigations described in this book edition and/or their generous help with various experimental measurements: Elisa Barbarese, Michael A. Davis, Donald C. Grant, Shih-Yieh Ho, Toyoko Imae, Inam U. Kureshi, Richard H. Simon, Candra Smith-Slatas, and Charles S. Springer. Finally other acknowledgments, in addition to those appearing in the chapters, include those in the first-edition preface following as well as permission for: using quoted material appearing on p. 13, Copyright © 1981 by the AAAS; and the reprinting of Figure 12.1 on p. 208, Copyright © 1991 by Sage Publications, Inc.

Joseph S. D'Arrigo

PREFACE - First Edition

    The occurrence of dilute gas-in-liquid emulsions [or coated microbubbles] in natural waters has been and continues to be a topic of great importance to workers in many fields of fundamental and engineering sciences. Specifically, the existence of long-lived gas microbubbles in fresh water, sea water, and other aqueous liquids including physiological fluids has been postulated and/or demonstrated by numerous investigators over the last three decades. In spite of this, no comprehensive account of the predominant physicochemical/biochemical mechanism by which such gas microbubbles (0.5-100 μm in diameter) are stabilized exists in the literature, and this book is intended to fill that gap. It begins with a review of the evidence for surfactant stabilization of the natural microbubbles commonly occurring in fresh water and sea water. The discussion continues with a description of microbubble experiments employing aqueous gels and soil extracts, then identifies many of the characteristic biochemical components of natural microbubble surfactant, and goes on to describe various geochemical, surface, and structural properties of the natural surfactant mixture. Combined with findings from physiological studies, this information is utilized for the successful production of relatively concentrated, significantly stable (hours to days), gas-in-liquid emulsions in artificial media, as described in the final chapters of the book.

    A detailed knowledge of the predominant physicochemical/biochemical mechanism by which gas microbubbles are stabilized in aqueous media is of practical importance to numerous and varied fields: acoustic and hydrodynamic cavitation, commercial oil recovery, hydraulic and ocean engineering, waste-water treatment, chemical oceanography, meteorology, marine biology, food technology, echocardiography, and the continual medical problem of decompression sickness. Many of these applications derive from the fact that persisting microbubbles affect the acoustical and mechanical characteristics of water, increasing attenuation, scattering ultrasonic energy, changing speed of propagation, and grossly reducing the tensile strength. Accordingly, the artificial enhancement of tensile strength in particular, i.e., through low-cost chemical destruction of the surfactant-stabilized microbubbles in water, is a desirable goal in order to improve the performance of devices normally limited in maximum output by cavitation, such as ship sonar, pumps, turbines, and propellers. The potential gains, such as the prevention of cavitation damage, obtaining a greater output from a given size or weight of equipment, i.e., an increase in return for a given economic investment, are all quite tempting. Separate, more fundamental considerations include the fact that microbubble populations, well known by marine biologists to exist in the upper ocean, can become attached to particles within the water column; this attachment affects the settling rates of marine detritus and, hence, has an impact on the ocean food chain. Also, bursting of bubble populations at the sea surface, with the concomitant production of a sea-salt aerosol and the ejection of organic material into the atmosphere, is of special interest to meteorologists, oceanographers, and environment specialists. As concerns industry, although adsorptive bubble separation has been used commercially for more than half a century (principally in froth flotation to separate minerals from ores), the related process of microflotation (requiring much smaller bubble sizes) was developed more recently for efficiently removing various colloidal pollutants from water and shows promise as a viable procedure for the treatment of water and waste waters. The process utilizes frothing agents, a potential area for further development, to promote the formation of microbubbles and these amphiphilic substances may contribute to the maintenance of a stable foam. Of more general interest to industry is the related, wider area of two-phase bubbly flows. Besides fundamental engineering interest, which aims at a deeper understanding of interphase momentum, heat, and mass transfer, there is a strong demand for practical information to optimize those systems in which two-phase flows occur, e.g., a motive liquid and an entrained gas. One common example of this type of chemical engineering operation, which might well be improved by the production of surfactant-stabilized gas-in-liquid emulsions, is the entrainment and pumping of corrosive fumes that are otherwise difficult to deal with. Moreover, artificially produced, surfactant-stabilized microbubbles would also have specific medical applications; for example, there is the immediate potential for developing longer-lasting and more uniform contrast agents for echocardiography, where injected air microbubbles have already been shown to travel with intracardiac velocities similar to those of red blood cells. Apart from echo-cardiography, another very promising clinical application of these nontoxic, synthetic microbubbles would be the ultrasonic monitoring of local blood flow in the abdomen (analogous to the current use of ordinary microbubbles to monitor myocardial perfusion). Such refined ultrasonic blood flow measurements, utilizing locally injected synthetic microbubbles, have the potential for providing better clinical detection of tumor neovascularization as well as any subtle changes in the normal vascularization patterns of organs neighboring abdominal masses. Hence, through the use of synthetic microbubbles, ultrasound may now provide much earlier diagnosis of abdominal masses; this early detection may well improve treatment of several classes of serious abdominal cancers, a notorious example being pancreatic cancer.

The underlying chemical principles covered in the chapters are presented in sufficient detail for this book to be useful to all interested readers with a working knowledge of chemistry, physics, and biology. Accordingly, the level of readership is intended to include graduate students, researchers, and professional people from widely varying fields. Furthermore, due to the many above-mentioned current and potential applications of stable gas-in-liquid emulsions, the appropriate readership of this book is likely to be found in industry, universities, and government laboratories alike.

Thanks are due to the following colleagues for their collaboration on some of the original investigations described in this book and/or their generous help with various experimental measurements: William Barker, J. Howard Bradbury, Kai-Fei Chang, Stephanie A. Ching, John F. Dunne, Richard J. Guillory, Brendon C. Hammer, Jacob N. Israelachvili, Kathleen M. Nellis, Barry W. Ninham, Noboru Oishi, Richard M. Pashley, Neil S. Reimer, P. Scott Rice, Cesareo Saiz-Jimenez, Kent Smith, Ourai Sutiwatananiti, and Linda Vaught. Finally other acknowledgments, in addition to those appearing in the chapters, include permission to use quoted material appearing on: p. 13, Copyright 1981 by the AAAS; p. 24, Copyright 1972 by the ASME; and pp. 7, 10, 16, and 93-94, Copyright 1975, 1978, 1978, and 1974, respectively, by the Pergamon Press, Ltd.

Joseph S. D'Arrigo

CONTENTS

Chapter 1

OCCURRENCE OF DILUTE GAS-IN-LIQUID EMULSIONS IN NATURAL WATERS

Dilute gas-in-liquid emulsions, namely, significantly stable (hours to days) suspensions of long-lived air microbubbles, in natural waters is a topic of great concern to workers in many fields of fundamental and engineering sciences. Specifically, the existence of stable gas microcavities or microbubbles in fresh water (ref. 1-29), sea water (ref. 5,15-17,27,30-33), and other aqueous liquids including physiological fluids (ref. 2,29,34,35) has been postulated and/or demonstrated by numerous investigators over the last half century. However, there is far less agreement in the literature as to the predominant physicochemical/biochemical mechanism by which such gas microbubbles, 0.5 ~ 100 μm in diameter, are stabilized. A detailed knowledge of this physicochemical/biochemical stabilization mechanism in aqueous media is of practical importance to numerous and varied fields: hydrodynamic (ref. 6,23,27) and acoustic (ref. 7,9-19,21,28) cavitation, hydraulic and ocean engineering (ref. 15-17,27), waste-water treatment (ref. 36), commercial oil recovery (ref. 37), chemical oceanography (ref. 38-42), meteorology (ref. 43-48), marine biology (ref. 49,50), food technology (ref. 51-53), and various medical applications such as echocardiology, decompression sickness (ref. 49,54-56; cf. Chapter 8) and, more recently, cancer diagnosis and treatment (see Chapters 8,12-14). Several of the above nonmedical applications (along with the underlying chemical and physical principles or considerations upon which they are based) are described individually in the section following.

1.1 PRACTICAL IMPORTANCE OF STABLE MICRO-BUBBLES

1.1.1 <u>Hydrodynamic cavitation, hydraulic and ocean engineering</u>

As Acosta and Parkin emphasize in a past review of this topic (ref. 27), hydrodynamicists do not need to be reminded that cavitation inception, with the pervasive role it occupies in naval architectural hydrodynamics, remains as a basic problem plaguing the worker in the laboratory and field alike. One of the more confusing aspects of this phenomenon has been its lack of repeatability between experiments carried out on similar test bodies in different test facilities (ref. 27) or even on different types of bodies in the same test facility (ref. 16,27).

For instance, appreciable tensile strength had occasionally been noted in water tunnels (ref. 16). (The tensile strength of a liquid is defined here as the minimum tensile stress in the liquid at which it ruptures or cavitates (ref. 57).) The inception of cavitation in these water tunnels has occurred at higher stress levels than ordinarily expected. Higher flow velocities or lower tunnel pressures than normal have been needed to produce cavitation about a test body. The tensile strength acts as if an additional static head were present in the system. In this case, appreciable tensile strength is undesirable in order to make for a uniformity of test results, and duplicate "prototype" conditions (ref. 16).

Alternately, the enhancement of tensile strength is a desirable goal in order to improve the performance of hydraulic devices normally limited in maximum output by cavitation, such as pumps, turbines, and propellers. The potential gains, such as the prevention of cavitation damage, obtaining a greater output from a given size or weight of equipment, i.e., an increase in return for a given economic investment, are all quite tempting (ref. 16).

The detailed work of Bernd (ref. 15-17) and other investigators has also shown that the tensile strength of water is set by the gas nuclei (i.e., microbubbles) present in the water. (Accordingly, the earlier-mentioned definition of the tensile strength of a liquid can be restated as the minimum tensile stress at which the gas nuclei in the liquid start to "explode". This property is also often referred to as the "cavitation susceptibility" (ref. 57).) Using specially constructed sonar transducers, the behavior of gas nuclei was followed by Bernd by measuring tensile strength. Surface films were found to form around the gas nuclei, retarding the acquisition of tensile strength by the water. To obtain a better understanding of these surface films, various waters were tested. Also surface films were created about gas nuclei (i.e.,microbubbles) from solutions of hydrocarbons (i.e., acyl lipids) and proteins in water (ref. 16).

Hence, in practical flow situations the water is not pure; gas bubbles and small impurities are embedded within the liquid. Small gas bubbles can stay in suspension for a long time, because the relative motion in an upward direction due to gravity is opposed by transport in the downwards direction by turbulent diffusion (ref. 57). These microbubbles are initially trapped in the liquid mostly by jet entrainment, cavitation, and/or strong turbulence at a gas/liquid (usually air/water) interface (ref. 58).

During the last four decades, measurements of weak nuclei (i.e., gas nuclei) in liquids have become especially important in view of their influence on cavitation inception (e.g., ref. 57-59). Understandably, the gas nuclei concentration is closely coupled with the free gas content of the liquid (ref. 58, 60). (A distinction is made commonly in the engineering literature between free gas content and dissolved gas content. The free gas content is that portion of gas which has the normal physical properties of bulk gas. In practical situations, the free gas concentration within the liquid is usually several orders of magnitude lower than the dissolved gas concentration (ref. 58).) Many investigators have developed instruments to detect this free gas content (ref. 58,60) and the "freestream" gas nuclei concentration associated with it (e.g., ref. 58-60).

The term "freestream nuclei" arises from the generally accepted view among engineers that cavitation inception occurs at nucleation sites which either originate from the test model surface or are present in the approaching flow. For a long time, "surface nuclei" attracted considerable attention. While it was shown that under certain circumstances the surface nuclei could exert a controlling influence on cavitation inception, it seemed evident on the other hand from the results of towing tank (i.e., model basin) tests that during "normal" cavitation testing freestream nuclei were more important (ref. 59). Hence, work by various investigators has been directed toward the development of equipment and techniques for determining freestream gas nuclei (i.e., microbubble) populations and attempting to relate them to cavitation inception (see ref. 57-60 for a selected summary of the past literature).

1.1.2 Acoustic cavitation

Ultrasonic cavitation has been demonstrated in aqueous systems for many years. Originally it was thought that the negative pressure in the sound wave actually tears the liquid apart; however, theoretical considerations make this highly improbable since they would lead to minimum necessary pressures of the order of at least several hundred atmospheres, not available in a sound wave. Accordingly, the presence of nucleation sites is evidently required (ref. 4). This latter idea was strongly reinforced when it was demonstrated by the experiments of Harvey (ref. 2) that undissolved gases were necessary for such nuclei. Harvey demonstrated that if water is subjected to 1,000 atm for 15 min and the pressure then removed, no cavitation could be produced for a considerable time. This could be explained by the effective destruction of the gas nuclei, the undissolved gas having been forced into solution by the high pressure (ref. 4). This still leaves two possibilities for the gas nuclei: undissolved gas adhering to solid particles, e.g., in a cleft or crevice, or free gas microbubbles.

Experiments to distinguish between these two possibilities have often involved measurements of ultrasonic attenuation (ref. 5,9,31,32). The popularity of this approach derives in part from the fact that small impurities in liquids, such as suspended particles, have negligible influence on attenuation in comparison with even a very small concentration of microbubbles (ref. 9). (Microbubbles, in contrast to solid particles, appreciably increase the compressibility of a liquid, introducing forms of viscous losses and nonreversible energy exchanges that do not exist in the case of solid particles.) It is therefore of considerable interest that all fresh tap water samples measured by Turner (ref. 9) showed substantial and persistent abnormal (ultrasonic) attenuation, amounting to a minimum of 44% over that of distilled water; it was concluded that this result stemmed from the presence of stabilized micron-sized bubbles.

Additional studies collectively suggest that there are approximately 10 times as many microbubbles in the sea as there are in fresh water, at least within the diameter range of 30-100 μm (ref. 50). It appears that some of these measured microbubbles in the ocean may not be generated continuously, but are persistent micro-bubbles with their buoyancy altered by attachment to particles, and with their dissolution rate greatly reduced by surfactant contamination (ref. 50).

Microbubbles in the upper ocean cause absorption and scattering of sound at frequencies near their resonant frequency, and change the sound-speed at frequencies near and below their resonant frequency. If present in sufficient numbers, they can therefore have a significant effect on acoustic propagation (and, as a practical example, sonar operations) (ref. 50). As relates to in situ acoustic measurements of oceanic microbubbles, it is the exaggerated effect of bubbles at resonance which allows the attribution of abnormal attenuation at a given frequency to the number of bubbles of a particular radius. Specifically, when a gas bubble in water is subjected to a sound at or near the bubble's natural frequency, it will very effectively absorb and scatter that energy. The effectiveness of the interaction can be described in terms of the total absorption cross section, the total scattering cross section, and their sum, the extinction cross section (ref. 32).

Accordingly, acoustic determinations of microbubble den-sities are based on the fact that resonating marine microbubbles have acoustical scattering and absorption cross sections that are of the order of 10^3 or 10^4 greater than their geometrical cross sections, and of the order of 10^{10}-10^{11} greater than those of a rigid sphere of the same radius, and that are still greater by far than those of common marine bodies of the same radius. This allows excess attenuation and backscatter of in situ experiments to be attributed to free microbubbles and biological bubbles rather than to particulate bodies (ref. 32). The justification for this interpretation has been given by Medwin (ref. 31,32). The identification of microbubbles has been validated by laboratory experiment and by the distinctive character of simultaneous measurement of backscatter and attenuation at sea (ref. 32). In addition, further analysis of the data from a variety of such past in situ studies has suggested that the larger bubbles (diameter > 120 µm) may be generated continuously by decaying matter on the sea floor, but that smaller bubbles (i.e., microbubbles with diameters < 120 µm) evidently persist indefinitely, once formed, because of surfactant coating and attachment to particles (ref. 50).

Finally, due to the surfactant coating surrounding stable

6

microbubbles in fresh water (ref. 8,15-17,25) and sea water (ref. 42,50), future acoustic determinations of bubble populations will need to be interpreted with regard for the effect of stabilizing surfaces on bubble resonance frequencies (ref. 42). This effect, if ignored, can result in and overestimation of bubble size from acoustic data (ref. 31,42).

1.1.3   Waste-water treatment: Microflotation

Although absorptive bubble separation has been used commercially for more than half a century (principally in froth flotation to separate minerals from ores), the related process of microflotation (ref. 36) was developed more recently for efficiently removing various colloidal pollutants from water and shows promise as a viable procedure for the treatment of water and waste waters. The technique can be distinguished from other flotation processes by requiring very low gas flow rates and extremely small bubbles. The adding of a collector agent, such as a long-chain fatty acid or amine, produces a stable surface phase and a thin layer of relatively dry foam upon which the floated material is collected. The process additionally utilizes frothing agents, a potential area for further development, to promote the formation of such micro-bubbles ($\sim$ 50 μm in diameter) and these amphiphilic substances may contribute to the maintenance of a stable foam. Owing to the small bubbles and low flow rates in microflotation, a relatively large gas/liquid interfacial area per unit of gas flow rate is produced with no severe agitation (ref. 36).

In contrast, most of the conventional foam separation tech-niques use large bubbles, requiring relatively high gas flow rates to generate sufficient interfacial area for adhesion of solid particles to bubbles. This causes turbulence at the foam/liquid boundary and, in order to prevent redispersion of floated particles, a rather tall foam column is required (ref. 36).

As summarized by Cassell et al. (ref. 36), the early work on microflotation described the removal of bacteria and algae from water (ref. 61, 62); thereafter, further studies demonstrated that B. cereus (ref. 63), colloidal constituents of tea (ref. 64), humic acid (ref. 65), colloidal silica (ref. 66), illite (ref. 67), titanium dioxide (ref. 68), and polystyrene latex (ref. 69) could be rapidly and efficiently removed from water by microflotation. The general vari-ety of dispersed materials which have been separated successfully by microflotation show this to be a nonselective method capable of removal of the mixtures of colloidal pollutants commonly found in natural and waste waters (ref. 70).

An important consideration with microflotation, however, is the fact that the work of Cassell et al. (ref. 36) shows rather conclusively that the optimum bubble size is ~ 50 μm. The efficiency of separation drops sharply when the bubbles are larger. Small bubble size is important in controlling the number of bubbles which may be attached to any given size of particle (ref. 36,71). Also, large bubbles could break the particle agglomerates formed during the flotation process, as well as produce turbulence which prevents the floating of particles. Finally, the high bubble densities achievable with microbubble production promotes good removal efficiency by increasing the probability of bubble-particle encounter and by making more bubbles available for attachment to each particle (ref. 36).

With these considerations in mind, the detailed work of Cassell et al. (ref. 36) was directed toward the optimization of the process parameters with special emphasis on bubble size as a function of surfactant (collector) and frother concentrations. Interestingly, after finding an extremely effective combination of these two classes of amphiphilic agents, they conclude "it is probable that exceedingly fine gas bubbles form which act as nuclei for the microbubbles upon introducing the gas through the frit" (ref. 36). Given the amphiphilic nature of the collector and frother agents used, it appears quite likely that these "exceedingly fine gas bubbles ... which act as nuclei" (cf. above) are, in actuality, film-stabilized microbubbles which soon thereafter grow into the somewhat larger microbubbles that participate effectively in the microflotation process. A more detailed chemical knowledge, and the resulting precise control, of the formation of such film-stabilized microbubbles (i.e., gas nuclei) may very well lead to even further improvements in, and reproducibility of, micro-flotation removal efficiencies in the future.

1.1.4 Marine biology, chemical oceanography

Many processes of oceanographic and biological significance have been attributed to the presence of bubbles in the near-surface region of the ocean. For example, Johnson and Cooke (ref. 33) point out that the role of bubbles has been described in such processes as gas exchange (ref. 72), organic particle formation (ref. 73,74), aerosol ejection (ref. 75), concentration of bacteria (ref. 76), and fractionation of organic and inorganic materials (ref. 77). In addition, if the bubbles are attached to particles within the water column, they will affect the settling rates of marine detritus and hence have an impact on the ocean food chain (ref.50).

Of the above-mentioned processes, organic particle formation has been and continues to be a topic of special importance for many marine biologists and chemical oceanographers. The early demonstration that dissolved organic carbon can be converted to particulate form by the bursting of bubbles (ref. 73,78) identified one mechanism whereby some proportion of the dissolved organic carbon in the oceans might be made available for utilization by organisms. As noted by Johnson and Cooke (ref. 41), this particulate material evidently serves as a substrate for bacteria (ref. 79) and is probably important in maintaining microzooplankton (ref. 80). The relative abundance of these particles in the ocean was demonstrated by Riley (ref. 79). He concluded, from in situ observations, that organic aggregates constituted a significant and often major part of the particulate matter in the sea.

Johnson and Cooke (ref. 41) additionally point out that since the early studies on bubble bursting and particle formation, several additional related processes have been identified as being of possible importance in converting dissolved organic carbon to particulate form. Wheeler (ref. 81) demonstrated that particles could be produced by compressing surface films, and Sheldon et al. (ref. 82) and Batoosingh et al. (ref. 83) concluded that organic particles form "spontaneously" in sea water. In studies of the rate of bubble dissolution, Liebermann (ref. 84) and Manley (ref. 8) observed that regardless of the treatment of the water, a residue always remains at the site where a bubble dissolves. Accordingly, bubble dissolution with subsequent aggregation of the adsorbed surface-active material was suggested by Blanchard (ref. 85) as another major means of organic particle formation in the ocean; this process was subsequently demonstrated in detail in the laboratory, using sea water, by Johnson (ref. 74) and later by Johnson and Cooke (ref. 41). Further work has confirmed that when a microbubble dissolves completely in sea water, the particle that invariably forms is composed of the surface-active material originally present on the microbubble at the air/water interface (ref. 42).

Hence, microbubble dissolution does consistently produce particles in sea water, but Johnson and Cooke (ref. 41) point out that evaluating the significance of the process in the oceans awaits a knowledge of organic particle composition. In view of the above-described arguments, it seems reasonable to assume that much of this needed knowledge can be obtained indirectly from careful study of the biochemical composition of the surfactant films surrounding natural microbubbles and, accordingly, provides yet another source of motivation for continuing research in this subject area. Once the necessary biochemical information on particle composition is obtained, along with further data on oceanic bubble populations, it is the feeling of some oceanographers (ref. 41) that the real significance of particle formation from bubble dissolution can be assessed in terms of the suitability of the particles as a foodstuff for animals, their contribution to the vertical flux of organic matter, and their tendency to concentrate pollutants and toxic materials.

1.1.5 Meteorology

During the past 40 years, meteorologists and environment specialists have become increasingly aware of the importance of natural sea bubble processes. Their importance to meteorology stems in part from the fact that surface-active organic material in the sea, mainly biological surfactants, tends to concentrate at the surface (ref. 85).

Evidence of this surface-active material on the sea can be seen in surface slicks or windrows (ref. 43). The composition of such surface films has received a great deal of attention (ref. 38-40,43-45,86-89). Most of this organic surfactant material must have its source in the dissolved organic matter in the sea (ref. 89,90). It is brought to the ocean surface by diffusion, by Langmuir circulations, and by the surfaces of air bubbles rising through the water. These bubbles are produced primarily by breaking waves (ref. 85); the majority of such bubbles are actually microbubbles with a diameter range of 50 to 120 μm (ref. 33) and a bubble size distribution peak at 100-120 μm diameter (ref. 33,91). These microbubbles not only carry surface-active material to the surface but, upon breaking, eject it into the air in the form of an aerosol (ref. 85). Accordingly, this process ejects not only the ocean surface constituents, but also the surface-active substances adsorbed onto the bubble during its rise through the organically rich upper segment of the sea (ref. 44).

In the past three decades, it has become clear that a rather large amount of surface-active organic material ends up in each tiny droplet ejected into the air by bursting bubbles. Some of these materials may reach concentrations in (or on) the droplets well over a thousand times their bulk concentrations in sea water (ref. 46,85,92). The water in the droplets that remain airborne eventually evaporates, leaving the nonvolatile materials to float around in the atmosphere (ref. 46) and ultimately settle out and, as a result, contribute appreciably to soil nutrients (ref. 93-95).

This natural process by which dissolved and/or particulate surface-active materials end up in the atmosphere has been modeled and studied in the laboratory. As summarized by Detwiler and Blanchard (ref. 46), tests in suspensions of bacteria (ref. 76,96,97), latex spheres (ref. 98), dyes (ref. 99), and in sea water and river water (ref. 96,100,101) have demonstrated successful transfer of all manner of surface-active material from the bulk fluid, or the bulk interface, to the droplets ejected when bubbles burst. (This situation can be pictured as an extension of the common industrial adsorptive-bubble-separation process (ref. 102) into a third dimension or phase -- the atmosphere.) Further laboratory tests with various tap waters, distilled waters, and salt solutions have shown that "no water sample was ever encountered that did not contain at least traces of surface-active material" (ref. 46).

The surfactant-rich droplets produced by ocean bubble bursting might very well also influence the growth and stability of sea fogs. As explained by Barger and Garrett (ref. 44), certain organic monomolecular films decrease the evaporation rate of small drops (ref. 103) and retard the growth of film-covered saline drops in atmospheres supersaturated with water vapor (ref. 104). Evaporation retardation becomes significant when the linear molecules of the organic (surfactant) film pack closely on a shrinking droplet surface (ref. 105-107). Because a natural surfactant film would concentrate at the surface of an evaporating drop, Goetz (ref. 108) has suggested that this process might stabilize sea fogs at low humidities and create a rather permanent haze of high organic content. Direct support for this idea can be found in the measurements on captured sea haze by Blanchard;

he determined that droplets rising from the sea in the surf zone carry a highly compressed surface-active film (ref. 43).

In addition to organic materials, the bursting of whitecap-induced bubbles at the surface of the sea is responsible for a vast sea-to-air transport of dissolved salts. This transport is estimated to be roughly 10^{10} tons per year (ref. 46,109). Accordingly, this transport has long been suggested (ref. 110) as the primary source of sea-salt nuclei. As a result, much attention has been focused on bubble spectra and how they produce the nuclei (ref. 33,48,75,99, 109,111-113). In addition, the surface-active material remaining on the sea-salt aerosol has given and will continue to provide marine meteorologists with a tracer which enables them to understand better the role of this aerosol in rain formation (ref. 85).

1.2   BACKGROUND OBSERVATIONS

1.2.1   Problems with the crevice model for bubble nuclei

Many investigators of the physics of bubble nucleation have assumed that bubble nuclei (i.e., gas nuclei) exist only in the crevices of hydrophobic surfaces (cf. "surface nuclei" in Section 1.1.1). Much of this work has been reviewed by Kunkle (ref. 114) who points out that these models are characterized by gas/liquid interfaces with infinitely large radii of curvature, i.e., flat surfaces. In practice this is accomplished by filling a crevice in a solid body with gas and by arranging the contact angles of the solid/liquid gas system in such a manner as to cause the curvature of the liquid/gas interface to vanish. The stabilization of gas in crevices is the mechanism responsible for the majority of the bubbles observed in common beverages supersaturated with gas, such as champagne and tonic water. Most of the bubbles in these systems do not originate in the liquid, but rather on the walls of the container (ref. 114). Harvey et al. (ref. 2) generalized this observation of bubble formation in crevices on container walls to nucleation within liquids by postulating that cavitation nuclei (i.e., gas nuclei) are small solid particles containing gas-filled crevices.

However, in addition to many problems in describing bubble nucleation in aqueous gels with the crevice model (see below, and Section 3.2), several fundamental observations suggest that this mechanism is not applicable to aqueous media in general (ref. 114).

Sirotyuk (ref. 25) found that the complete removal of solid particles from a sample of water increased the tensile strength by at most 30 percent, indicating that most of the gas nuclei present in high purity water are not associated with solid particles. Bernd (ref. 15,16) observed that gas phases stabilized in crevices are not usually truly stable, but instead tend to dissolve slowly. This instability is due to imperfections in the geometry of the liquid/gas interface, which is almost never exactly flat (ref. 114). Medwin (ref. 31,32) attributed the excess ultrasonic attenuation and backscatter measured in his ocean experiments to free micro-bubbles rather than to particulate bodies; this distinction was based on the fact that marine microbubbles in resonance, but prior to ultrasonic cavitation (ref. 4), have acoustical scattering and absorption cross sections that are several orders of magnitude greater than those of particulate bodies (see Section 1.1.2).

Cavitation experiments have produced two additional results which indicate that most gas nuclei are not of the crevice type. In filtration experiments with gelatin reported in Kunkle's study (ref. 114), it was found that the size of a nucleus is strongly correlated with the critical supersaturation pressure required for bubble formation. In the context of the crevice model, this result would mandate that crevices of a given size and shape are associated uniquely with particles of a single size. Because it seems likely that the types of clefts occurring in small particles would also occur in larger ones, this experimental result argues against the applicability of the crevice model to gelatin (ref. 114). In addition, both Strasberg (ref. 7) and Bernd (ref. 15,16) noted that the threshold pressure for ultrasonic cavitation in water can be increased by allowing the water to stand quietly for many hours, and Bernd showed that this increase can be consistently explained by the rise of the largest nuclei to the surface of the water. Since solid particles generally will not float, these experiments again indicate that gas nuclei are, at a minimum, not exclusively of the crevice variety (ref. 114).

In summary, nuclear models of the crevice type consist essentially of gas phases stabilized in crevices in solid particles. While the crevice hypothesis represents a viable nuclear model, none of the existing mathematical treatments make predictions that are supported by the above-mentioned gelatin experiments (ref. 114). In addition to these problems with the mathematical development of this model, various experimental studies (cited above) show little or no correlation between the cavitation susceptibility of a liquid and the number of solid impurities in the sample (see also Section 3.2). The gas nuclei found within aqueous gels, fresh water, and most aqueous liquids are, in all probability, mostly not of this type.

1.2.2 Reduction of gaseous diffusion across the air/water interface by selected surfactant monolayers

Minute bubbles stabilized by encapsulation in an organic film have long been invoked to explain low thresholds for ultrasonic cavitation in water (ref. 4). Small gas bubbles can stay in suspension for a long time, because the relative motion in an upward direction due to gravity is opposed by transport in the downwards direction by turbulent diffusion. These microbubbles are initially trapped in the liquid mostly by jet entrainment, cavitation, and/or strong turbulence at the air/water interface (ref. 58). Experimental work by Manley (ref. 8) on bubble dissolution in distilled water revealed that the effective diffusion coefficient for air through bubble walls is reduced at very small bubble diameters. This observation, combined with results from simple chemical additions, was interpreted as an indication that an organic skin of impurity surrounds each microbubble (ref. 8). Similar and more recent experiments by Johnson and Cooke (ref. 42) on bubble dissolution in sea water have demonstrated that many such bubbles did not dissolve completely; instead, these small bubbles "stopped decreasing in size abruptly, sometimes becoming slightly aspherical, and remained as microbubbles apparently stabilized by films compressed during dissolution. ... It is obvious that some mechanical or diffusion-limiting effect that resists further bubble dissolution is activated during bubble contraction and the attendant compression of encapsulating surface films" (ref. 42).

Actual measurements of reduced gaseous diffusion across certain compressed surfactant monolayers, at an air/water interface, have been reported in detail by numerous investigators in the past (ref. 115-120). For such measurements to be completely trustworthy, it is first necessary to eliminate convection in the bulk (aqueous) phase, since the monolayer can reduce the rate of gas absorption by reducing convection at the surface, and this has a greater effect than a "diffusion" barrier (ref. 116). The problem of convection has been treated in detail (ref. 115,116), and fairly conclusive proof was offered, in the case of carbon dioxide absorption by alkaline buffers, that convection was essentially absent. Briefly, the calculated (advancing front diffusion) absorption rate without monolayers agreed with the measured rate using _permeable_ monolayers (as opposed to relatively _impermeable_, insoluble monolayers); the permeable monolayers used (e.g., bovine serum albumin), effective only in cases of convection, were ineffective in these experiments. In addition, the values calculated for relatively impermeable monolayers (concerning CO_2 permeation) were independent of the absorbing solutions used. These demonstrations allow the data to be treated as "diffusion" measurements as opposed to convection measurements (ref. 116).

Another potential pitfall to be avoided with such monolayer measurements, as pointed out by Blank (ref. 116), can be found in an early study by Linton and Sutherland (ref. 121). They reported on the diffusion of oxygen through a monolayer, the gas passing from the vapor phase into the liquid phase. They found no significant effect for a fatty alcohol monolayer under quiescent conditions, since there is a great resistance to oxygen absorption in the water itself. Their result demonstrates another important consideration in such monolayer experiments, namely, that "the permeability of a monolayer to gases must be several orders of magnitude smaller than that of water in order to affect the observed gas absorption rate. If one removes the gas once it has passed through the film (by a chemical reaction in the aqueous phase), this increases the relative permeability of the water and makes it easier to detect an effect for a monolayer" (ref. 116).

Accordingly, Blank (ref. 115,116) reported an improved method for determining monolayer permeability to gases by means of accelerated gas absorption. In this method, the monolayers were spread at the gas/water interface, and the direction of flow was from the gas phase into the aqueous subphase. Since convection was shown to be absent, the permeability of several monolayers to a variety of gases could be determined.

The initial experiments (ref. 115,116) utilized 0.1 M carbonate buffer as the absorbing solution and CO_2 as the gas. (A slight retardation effect for a monolayer could be observed upon absorption into pure water, but since CO_2 reacts with water causing a change of pH, alkaline buffers were used, and this increased the overall rate of gas absorption as well.) The relatively impermeable monolayers, consisting of (saturated, linear) fatty alcohols or acids, had a large effect on the gas absorption rate. The experiments performed with permeable monolayers of bovine serum albumin showed no difference from the absorption rate without a film. Experiments with other alkaline buffers (glycine (pH 9.8), Tris (pH 8.3), and hemoglobin (pH 10.5)) gave similar results (ref. 116).

Additional experiments (ref. 116) employed a 0.2 M $Na_2S_2O_4$, 1.0 M NaOH solution as the aqueous subphase and oxygen as the gas. Because of the relative insolubility of oxygen, the alkaline $Na_2S_2O_4$ solutions were used to increase the O_2 uptake rate in the liquid and, thereby, enhance the relative effect of a monolayer. The absorbing solutions needed to be alkaline, since gases are evolved from the reaction of oxygen with $Na_2S_2O_4$ at lower pH. The results for this case were similar to those obtained for carbon dioxide with regard to the relative effects of the various (insoluble and rather impermeable) monolayers used, supporting the belief that only a "diffusion" barrier was involved in these experiments. (The only difference was in the case of C_{18} acid where the absorption of CO_2 (at pH 10) was retarded but that of O_2 (at pH 13) was unaffected. However, there is reason to believe that at the higher pH, the dissociation of the COOH group caused considerable expansion of the monolayer and hence greater permeability. The fatty alcohol monolayers were not affected at this higher pH (ref. 116).)

Other experiments (ref. 116) involved the absorption of nitrous oxide into pure water. (The absorption rate, into pure water or into aqueous solutions, was sufficiently large to enable detection of monolayer effects.) Nitrous oxide, which is about as soluble as carbon dioxide, has the advantage of not reacting chemically with water, and this enabled the study of monolayer permeability on low pH subphases. Accordingly, similar results were obtained for nitrous oxide absorption into a 0.1 M HCl solution. The relative effects for the different (saturated acyl lipid) monolayers were as in the cases of oxygen and carbon dioxide, and the magnitudes were about the same. This allows one to assume the absence of convection effects, at least during the beginning of the absorption where the relevant data were taken (ref. 116).

The above-described results obtained by Blank for carbon

dioxide, oxygen, and nitrous oxide indicate that (relatively impermeable) monolayers are "diffusion" barriers to gas absorption and that the effects for these various (insoluble) monolayers are approximately independent of the gases studied or the various solutions used to absorb the gases. This represents a strong cross check on the experimental procedure and on the data themselves. The semiquantitative conclusions drawn by Blank (ref. 116) from the resulting calculated permeabilities for the three gases were that: a) the values are approximately the same; b) the values are approximately two orders of magnitude lower than the permeability to water vapor; and c) for the same surfactant polar group, the values decrease with increasing (saturated, linear) carbon chain length.

Similar experimental results and conclusions have been reported by Hawke and Alexander (ref. 117) and by Hawke and Parts (ref. 118). These authors also point out that these saturated acyl lipid monolayers need to be in the close-packed (i.e., laterally compressed) condition before significant reduction in gaseous diffusion is observed (ref. 117,118; see also Sections 1.3.1 and 7.6).

1.3 DEMONSTRATION OF FILM-STABILIZED MICRO-BUBBLES IN FRESH WATER

1.3.1 Acoustical measurements

In line with discussions included in previous sections, ultrasonic experiments carried out on fresh water by different investigators indicate that the stabilization of gas microbubbles, acting as gas nuclei for ultrasonic cavitation, is always attributable to the presence of surface-active substances in the water (ref. 15-17,25). As a starting point, one should consider that laboratory tests with various tap waters, distilled waters, and salt solutions have shown that "no water sample was ever encountered that did not contain at least traces of surface-active material" (ref. 46). Sirotyuk (ref. 25) estimates that the content of surface-active substances in ordinary distilled water amounts to $\sim 10^{-7}$ mole/liter, and in tap water it is $\sim 10^{-6}$ mole/liter or higher. These values indicate the appreciable content of such substances in both cases (ref. 122), although they differ by roughly an order of magnitude in absolute value. It is essentially impossible to completely remove these surface-active substances by any of the most diverse water-purification techniques (ref. 25,122). Hence, gas bubbles formed in the liquid as a result of its motion can be stabilized by the minute quantities of surface-active substances that are always present in water. Accordingly, Sirotyuk (ref. 25) predicted that in specially purified water containing only vestiges of surface-active substances, the number of gas bubbles and their diameters should decrease and the tensile strength (or cavitation threshold) of the water increase (ref. 25,123).

For the preparation of this specially purified water, Sirotyuk (ref. 25) first filled his experimental chamber with an alkaline solution of $KMnO_4$, which was boiled for 10 hr with the streaming of oxygen through it. Distillation was carried out, on the other hand, under a helium atmosphere. The content of surface-active substances in the water was estimated, by adsorption polarographic analysis (ref. 122,124), to be close to 10^{-8} mole/liter. The actual measurements of the cavitation threshold of this high-purity water were carried out in a specially built apparatus which was completely hermetically sealed. Consequently, the water in the experimental apparatus was kept from contact with the atmosphere right up to its admission into the reaction vessel in which the cavitation threshold was determined. Upon testing this high-purity water, Sirotyuk found, in line with his above-described expectation, that the cavitation threshold was significantly elevated. Specifically, it was clear that the reduction in the content of surface-active substances to $\sim 2 \times 10^{-8}$ mole/liter caused a marked increase in the cavitation threshold of the water to ~ 7.5 atm (ref. 25). (At the same time, the fact that the cavitation threshold did not even begin to approach the theoretical molecular strength of absolutely pure water, which is on the order of 10^4 atm (ref. 25), confirms a related experimental observation by Bernd (ref. 17); namely, a surfactant film can successfully form around a gas nucleus (i.e., microbubble), prior to and preventing dissolution of the microbubble, with a concentration of film-forming materials as low as 10^{-9} mole/liter (ref. 17).)

There is other acoustical evidence to support the belief of Sirotyuk and other investigators that stable microbubbles serve as cavitation nuclei in fresh water. As noted by Sirotyuk (ref. 25), numerous experiments have disclosed that the cavitation threshold of water is increased by degassing of the liquid or by the preliminary application of static pressure. This finding is consistent with the belief that gas microbubbles are always present in water, acting as primary cavitation nuclei (ref. 25,124). Furthermore, this finding agrees with earlier-mentioned observations (see Section 1.1.1) that (hydrodynamic) cavitation inception, and the gas nuclei population associated with it, is closely coupled with the free gas content of the liquid (ref. 58,60). (In practical situations, the free gas concentration within the liquid is usually several orders of magnitude lower than the dissolved gas concentration (ref. 58,125).)

Other acoustical measurements on degassed or pressure-treated fresh water, similar to those mentioned by Sirotyuk, were performed in some detail by Iyengar and Richardson (ref. 5) more than a decade earlier. As these authors point out concerning degassed water, absorption of sound waves in such water is small at frequencies below 2 MHz. At these frequencies, the presence of small amounts of air entrained in the water would considerably increase the absorption. This excess absorption was measured by Iyengar and Richardson using a reverberation technique. In this method, measurement is made of the rate of decay of a diffuse sound field established in the water. The attenuation is caused by absorption in the water itself and also by energy losses associated with the vessel. (The vessel losses were ascertained by using a liquid for which the coefficient of absorption of sound was known.) The reverberation vessel was first filled with degassed water which had been kept standing for several days and the reverberation times were determined at a number of frequencies. Then tap water, or some other water specimen, was substituted. The difference in the reciprocal of reverberation time, defined by Iyengar and Richardson as the "decay factor", is taken as a measure of the sound absorption due to air bubbles. Results for tap water showed that the "decay factor" at any given frequency decreased with time. The authors interpreted their data to mean that the larger bubbles disappear (via flotation) first, then the smaller ones, and so the decay constant continues to fall, until finally after 20 hours it becomes almost identical with that of degassed water. Accordingly, by applying Stoke's law (ref. 126), Iyengar and Richardson arrived at an estimate of the size (diameter) of the microbubbles which can reach the surface at the end of this period; it was of the order of a micron (ref. 5).

Separate acoustical measurements performed in an ultrasonic cavitation tank, involving pressure-treated tap water samples, were also reported by Iyengar and Richardson (ref. 5). The tank was filled with degassed water so that (gaseous) cavitation would not take place even at maximum sound intensities (40 atm). A clean thin-walled glass bulb was filled with tap water, introduced into the tank at the acoustic focus, and the cavitation threshold of the sample observed. The bulb with the water sample was then transferred to a pressure vessel in which hydrostatic pressures up to 10,000 lb/in^2 could be applied with the aid of a hydraulic pressure intensifier. (In the experiments, pressures from 100 lb/in^2 upwards were applied for 3 to 5 min.) The pressure was then released and the bulb with the contents was again introduced into the tank at the acoustic focus. The thresholds observed in this manner showed that there was a rapid increase of threshold with the degree of undersaturation. The time-effect of pressurization was also studied, both at 100 lb/in^2 and 300 lb/in^2, and it was found that the cavitation threshold of the water increased asymtotically with time. In addition, the ultrasonic reverberation time of the water was determined before and after pressurization. A decrease in acoustical absorption was found and interpreted by Iyengar and Richardson to represent the entrained air (i.e., gas nuclei) going into solution.

All the above-described properties of gas microbubbles in fresh water including the effects of pressure, their flotation and/or dissolution with time, and their stabilization by organic films are brought together and reconfirmed in the detailed work of Bernd (ref. 15-17). This investigator proposed that in a body of water, the tensile strength (i.e., cavitation threshold or susceptibility) should increase with depth, due to several actions. Generally speaking, the tensile strength of the water is inversely proportional to the maximum size of gas nuclei (microbubbles) present. Large micro-bubbles, that prevent appreciable tensile strength from occurring, rise rapidly to the surface; small microbubbles rise slowly and so remain behind. However, small microbubbles tend to dissolve. The high ratio of interface surface to gas volume, and internal pressure created within the bubble by surface tension, favor the dissolving of small microbubbles. Dissolving is also promoted by increasing depth (i.e., increasing hydrostatic pressure). Hence, microbubble size decreases with increasing depth because of the relative rates of rise, and because of dissolving. To test this hypothesis, Bernd (ref. 16) performed a variety of tensile strength tests in fresh water. Specifically, Bernd measured the tensile strength obtained versus depth in two fresh water lakes, using specially constructed sonar transducers to stress the water. The tensile strength did in fact increase with depth. For example, a 6.7-db increase in power level above the "normal" inception of cavitation (i.e., near the lake surface) was obtained at a 39.4-foot depth before cavitation took place. Thus, a 36.1-psi tensile strength was obtained at a 39.4-foot depth, and the transducer brought about cavitation as if it were at a 123.0-foot depth instead of 39.4 feet. Accordingly, such experiments show that appreciable tensile strength exists in natural water at moderate depth (see Fig. 1 in ref. 16).

Despite this correlation with water depth, Bernd (ref. 15,16) re-emphasized the common argument that these microbubbles or gas nuclei should theoretically dissolve completely within seconds to a few minutes but, as shown by numerous experiments, often do not. This situation retards the acquisition of tensile strength, and limits the amount of tensile strength available in the water. Bernd (ref. 16) described this phenomenon to be a consequence of surface films that gather about the gas nuclei from small amounts of organic materials dissolved in the water. He presents much evidence to show that various fresh waters tested varied widely in their tendency to form these surface films, as evidenced by variations of more than 100:1 in the dissolving rate of the gas nuclei contained in the water. Waters that did not produce surface films dissolved gas nuclei quickly. "For example, water from the David Taylor Model Basin's test tunnels acquired tensile strength quickly after being cavitated. Here, one would attribute the lack of surface films to the fact that chlorinated water is used in the tunnels. In comparison, the 'prototype' sea water and lake water tested were approximately an order of magnitude slower in dissolving. When heavy organic growths were present in water, much slower dissolving -- or none at all -- resulted" (ref. 16). In additional experiments by Bernd, surface films were deliberately produced around gas nuclei "by dissolving small quantities of hydrocarbons or proteins in water. It was found possible to obtain faster or slower dissolving than the 'prototype' norm, and largely duplicate the action of various natural waters" (ref. 16). Such surface films around gas nuclei are produced by trace organic materials possessing solubilities on the order of 10 to 100 parts per million. These materials are surfactants, i.e., they are able to migrate through the liquid to the surface of the microbubble and form a monolayer. Thus a low surfactant concentration in solution is brought to a high concentration within the monolayer at the microbubble surface. In some natural waters, the monolayer itself may also become surrounded by additional structure-building materials (ref. 16).

Furthermore, three classes of microbubble dissolution were obtained and analyzed in a subsequent study by Bernd (ref. 17) of surface-film effects: 1) a rapid rate of dissolving when a surface film was absent; 2) a rapid rate of dissolving when a surface film was present but uncompressed; 3) a slow rate of dissolving when a surface film is present and compressed. The degree of slow dissolution, observed with microbubbles having compressed surface films, was found to depend heavily on the nature of the surfactants contained in the film. In water tunnel tests utilizing film-forming additives, measurements revealed saturated, linear acyl lipids to be especially effective in preventing microbubble dissolution (ref. 17).

From such microbubble-dissolution measurements, Bernd (ref. 16,17) outlined a physical model to explain much of the dynamic behavior of film-stabilized microbubbles. One problematic aspect of this dynamic behavior involved the question of how a gas nucleus can be surrounded by a relatively impermeable film and yet subsequently act to produce cavitation when a gas/water interface is needed to initiate cavitation. Bernd (ref. 16) explains that if the stabilized gas microbubble enters a low-pressure area, the gas within the microbubble will attempt to expand. The surfactant film may also elastically attempt to expand. The surfactant film will then be expanded until essentially the surface tension of the water alone acts to contract the microbubble, since the protective "shell" no longer acts. The film has either been ruptured upon expansion, or it has expanded until it is ineffectual. Thus the microbubble (i.e., gas nucleus) should be capable of expanding to form a cavitation void or acquire additional gas in the form of water vapor or from surrounding dissolved gas. In addition, Bernd points out that it is reasonable to expect a gas microbubble to acquire such an effective monomolecular surface film. The characteristics of many surface films of biological surfactants observed at flat gas/water interfaces (e.g., in Langmuir-trough experiments) imply that the necessary properties for a stabilized gas nucleus capable of initiating cavitation are obtainable by a surfactant film at a microbubble surface (cf. Section 1.2.2). For example, investigations on the behavior of surfactant films at a gas/water interface have shown that some monomolecular films can be highly compressed (on the order of 50:1) as the interface is contracted, and then subsequently re-expand to their original condition as the interfacial area is expanded. Other surface films may, however, change their physical structure upon compression, possibly acquire a permanent set, slowly redissolve in the water, or yield slowly in a viscous fashion; thus hysteresis may be exhibited for various reasons (ref. 16).

1.3.2 Light-scattering measurements

In addition to acoustical methods, which take advantage of the fact that gas nuclei (i.e., stable microbubbles) are elastic bodies and thus absorb sound energy (ref. 4,5,9,25,26,31,32,50), another class of methods for detecting these gas microbubbles that has been employed repeatedly is based on their optical behavior. Specifically, most of these optical methods involve detection of these long-lived microbubbles in water from the light scattered by them (ref. 5,26,59,60,127).

From their early experience with long-lived microbubbles using the acoustical methods of detection, Iyengar and Richardson (ref. 5) recognized that some sorting of microbubble sizes with time occurs from the differential rate of rise of such microbubbles in the liquid (see Section 1.3.1); this finding suggested to the authors a method for the delineation of microbubble size in fresh water based on Mie scattering (ref. 5), which could then be compared to their earlier calculations utilizing Stoke's law. According to Mie theory, when a microbubble passes through a control volume of high light intensity, it will scatter light, the intensity of which is proportional to the microbubble's geometrical cross section within certain limits (ref. 5,26,128). When the microbubble diameter is much smaller than the wavelength of incident light, the intensity of scattered light is symmetrical about the direction of the incident light, as well as about a perpendicular direction (ref. 5). However, as microbubble diameter approaches the wavelength of incident light, more light is scattered in the forward than in the backward direction (ref. 5,129). Accordingly, Iyengar and Richardson (ref. 5) measured the variation of intensity of scattered light (λ = 0.408 µm), as a function of the angle between the directions of the incident and the scattered light, at first in freshly aerated water. The resulting curve (i.e., intensity scatter envelop) was repeatedly observed to be disymmetrical about an axis perpendicular to the light path, with the scattering in the forward direction being much larger; the minimum of the envelop was thus displaced in the backward direction. Interestingly, with increasing time (i.e., at 1 and 10 hours after aeration), the magnitude of this backward displacement was found to decrease progressively (along with a parallel decrease in the relative scattering in the forward direction). However, the symmetry which is characteristic of degassed water was never attained even after prolonged standing. Iyengar and Richardson (ref. 5) explain this sequence of results by pointing out that depending on the number of the largest microbubbles present, the minimum will be shifted from the 90° position in the backward direction. After the large microbubbles have risen, the smaller microbubbles which are left behind, while scattering less light, will give rise to a minimum which is nearer the 90° position. This process will continue until the smallest microbubbles are left in the liquid. Finally, to estimate the size of these remaining microbubbles, Iyengar and Richardson compared the intensity scatter envelop obtained for long-standing fresh water with the predictions of Mie theory. From the position of the minimum and the amount of asymmetry in the experimental curve, the calculated microbubble size in long-standing water was 0.8 µm (ref. 5). This result is in good agreement with their earlier-described finding (see Section 1.3.1) from acoustical tests, and the application of Stoke's law, that the diameter of the microbubbles which can reach the surface in long-standing (i.e., 20 hr) tap water is on the order of a micron (ref. 5).

More recent light-scattering studies (ref. 26,59,60) of microbubble populations in fresh water, using laser-light sources, have yielded very similar results. For example, Keller's laser-scattered-light technique (ref. 26) provided precise measurements of the size and number of freestream gas nuclei (i.e., long-lived microbubbles) in a cavitation tunnel; from microbubble spectra recorded within the diameter range of 7-17 μm, it was repeatedly determined that "merely letting the water rest" for a period of one hour was sufficient to cause significant alteration (i.e., decreased mean diameter and microbubble number) in the recorded microbubble spectrum, which was paralleled by an increase in cavitation threshold of 50% or more (ref. 26). In addition, Keller found that "In contaminated water, more or larger nuclei can stabilize than can in clean water; as a result, the cavitation starts earlier and more abruptly" (ref. 26). From his various measurements, Keller concluded "The scattered light must predominantly originate from the bubbles and to a much lesser extent from [particulate] contamination, otherwise the nucleus spectrum would not change so much with time. In order to prove that the change of the nucleus spectrum was not caused by the settling of the contamination particles, the following test was carried through: Water which had rested for a long time was pressurized, and the nucleus spectrum was measured. Afterwards the water was circulated. The velocity was such that, as a result of the [pre-existing] high static pressure, cavitation could not occur; however, the settled contamination particles would definitely have been stirred up. A consecutive measurement of the nucleus spectrum [without water circulation] did not produce any change in the spectrum" (ref. 26). Keller adds that "even the comparison of spectrums obtained for the same water at different pressures demonstrates that the scattered light originated predominantly from the bubbles. If one subjects the water to different pressures and records the associated spectrums, one obtains, with increasing pressure, a decrease of the nuclei size -- a result which can be explained only by a decrease of the nuclei as a result of [gas] diffusion and compression" (ref. 26).

1.3.3 Gas-diffusion experiments

As noted several times in this chapter (see Sections 1.2.2 and 1.3.1), various experiments in different types of fresh water indicate that the effective diffusion coefficient of gas through bubble walls is significantly reduced at very small bubble diameters. One example mentioned only briefly earlier, and useful to examine further, is Manley's study (ref. 8) of the change in diameters of small bubbles with time in distilled water. Using a microscope to make his measurements, Manley found that for diminishing bubbles in undersaturated water, the square of the diameter (D) of the bubble no longer changes in proportion with time (t) when the bubble is very small. Specifically, the change of D^2 with t for six bubbles of different initial volumes was recorded in undersaturated water; the calculated diffusion coefficient (ref. 8,84,130) of air out of these bubbles was found to be reduced when the diameter of each bubble dropped under 100 μm (ref.8). Additional experiments showed that there was very little change in the diffusion rate out of such microbubbles in water which had recently been well degassed. Consequently, Manley ruled out the possibility that the slow diffusion rate was primarily the result of saturated layers of dissolved gas around the microbubble. Further experiments revealed that the addition of a wetting agent (Lissapol N) to the water actually eliminated this reduction in the diffusion rate of air into or out of such microbubbles in distilled water. From this and the above results, Manley concluded that the reduced gaseous diffusion across a microbubble interface is due to "an organic skin of impurity" surrounding the microbubble (ref. 8).

1.4   DEMONSTRATION   OF   FILM-STABILIZED   MICRO-BUBBLES IN SEA WATER

1.4.1   Acoustical measurements

Iyengar and Richardson (ref. 5) recognized more than four decades ago that the cavitation threshold of sea water is routinely significantly lower than that of most fresh waters. In one experiment, they collected two samples of sea water near the shore, on different days, and subjected them to ultrasonic cavitation. The samples showed a low cavitation threshold, lower than that recorded for tap water. The sea water samples were then allowed to remain undisturbed for a few days longer. Once again, at the end of the extended quiescent period, they showed the same relatively low threshold. On examination of the air content of the samples, they showed undersaturation, the actual percentage saturation being 85%. Thus it appeared that they contained some air in undissolved form, probably entrained by the action of waves (or produced by organisms near the surface). The sea water samples were thereafter pressurized at 100 and 200 lb/in^2. At the lower of these two pressures, their cavitation threshold was increased to 15 atm, while at the higher pressure no cavitation could be produced (ref. 5).

The above results are readily understandable in view of the demonstrated fact, referred to in a previous section (see Section 1.1.2), that the average microbubble concentration (as well as average bubble size) in sea water is significantly higher than that in fresh waters (ref. 50,60); more specifically, from acoustical measurements of microbubbles in the diameter range of 30-100 µm, it is reported that there are approximately 10 times as many microbubbles in the sea as there are typically in fresh water (ref. 50,131). This estimate is based on the observation that the long-lived microbubbles measured in fresh water by various investigators (e.g., ref. 9,12,22) had diameters below about 100 µm. The microbubble size distribution from these reports were then compared by Medwin (ref. 131) and Mulhearn (ref. 50) with the corresponding oceanic data of Medwin (ref. 31,131). (As one example of such oceanic data, the concentration of microbubbles, with diameters between 20 and 120 µm, measured by Medwin in August 1975 in 40 m of water in Monterey Bay (ref. 32) was approximately 10^4/liter.) In addition to comparing microbubble densities in sea water versus fresh water, Mulhearn provides evidence which indicates that in sea water these microbubbles (ranging in diameter up to 120 µm) persist indefinitely, once formed, because of surfactant coating (preventing their dissolution) and alteration of their buoyancy by probable attachment to particles (ref. 50).

As mentioned earlier (see Section 1.1.2), acoustic determinations of actual microbubble populations in the ocean are based on the fact that resonating marine microbubbles have acoustical scattering and absorption cross sections that are of the order of 10^3 or 10^4 greater than their geometrical cross sections, and of the order of 10^{10}-10^{11} greater than those of a rigid sphere of the same radius, and that are still greater by far than those of common marine bodies of the same radius (ref. 32). Accordingly, the distinctive and exaggerated acoustical cross sections of a single resonant microbubble permit it to be selectively identified in the presence of particulate matter or non-bubble-carrying marine animals (ref. 31).

1.4.2 Light-scattering measurements

Besides the earlier-mentioned light-scattering studies of microbubble populations in fresh water (ref. 26,59), Weitendorf (ref. 60,132) has also conducted a similar laser-scattered-light study in sea water in order to obtain precise measurements of the size and concentration of long-lived, freestream gas nuclei (i.e., microbubbles) in the open ocean. The measurements were carried out on board of a transoceanic container ship, while traveling at various ship speeds, using optical instruments installed in the ship's hull. The optical recordings themselves were made within an optically defined control-volume of 0.98 mm^2. These "full scale" investigations clearly indicated that "in the free sea always a large number of [gas] nuclei is present" (ref. 60), i.e., significantly higher than the microbubble concentrations found by Weitendorf in fresh water during model tests within a cavitation tunnel (ref. 60).

The actual distributions of microbubble diameters in sea water, measured by Weitendorf in the optically defined control volume, ranged between 20 and 117 μm. Within this diameter range, the usual number of measured microbubbles per cm^3 was of the order of 10 to 100, yielding an approximate microbubble concentration of 10^4-10^5/liter for these ocean experiments (ref. 60,132). In addition, the maximum of this gas nuclei distribution occurred in the diameter range of 20-40 μm (for sea water samples taken both inside and outside the seaway, i.e., shipping lane). Weitendorf points out that this diameter range (20-40 μm) consists almost exclusively of microbubbles (ref. 60,132). (Contrariwise, the range of nuclei with diameters of 20 μm and below, which as noted above were disregarded in Weitendorf's experiments, consists of suspended particles as well as microbubbles (ref. 60,133,134).)

1.4.3 Photographic identification

Separate from his earlier-described acoustical measurements, Medwin (ref. 31) conducted a direct photographic determination of microbubble populations in coastal ocean waters. His results indicated the presence of several million microbubbles per cubic meter (i.e., >10^3/liter) within the diameter range of 20 to 100 μm, both at the surface and at a depth of 10 feet in Monterey Marina (ref. 31,33). However, as a result of his technique of illumination, Medwin suggested that his count probably included many non-bubbles (ref. 31).

To extend this type of microbubble study, Johnson and

Cooke (ref. 33) subsequently developed an improved photographic apparatus for the direct examination of oceanic bubble populations. The apparatus consisted of an electrically driven camera mounted in a pressure housing. Three small strobe lamps equally spaced around the perimeter and slightly behind the zone of focus of the camera provided the necessary illumination and produced images on the film corresponding to three points (of specular reflection) on the spherical surface of each bubble in the field of view. Bubbles thus appeared in the photographs as easily identifiable groups of three spots (hereafter referred to as "image dots") that lay at the vertices of equilateral triangles. From the dimensions of any of these triangles, the diameter of the associated bubble could be readily determined by Johnson and Cooke. The problem, mentioned by Medwin in his early study (ref. 31), of an inadequate sample volume in this photographic approach could only be resolved by choosing the proper magnification, a strategy that was limited by the necessity of resolving bubbles smaller than 40 μm in diameter. The film and developer combination used gave measurable images of bubbles of less than 40-μm diameter when the image magnification was 0.33X. Sample volumes were maximized by Johnson and Cooke by measuring all bubble images for which the image dots did not intersect. When a bubble is increasingly out of focus, its three image dots become larger, but the distance between the centers of the dots remains very nearly constant and consequently allows the bubble diameter to be readily calculated. By using this approach, the depth of field becomes a function of bubble size and is therefore determinable by calibration. The calibration was performed by submerging the camera housing in a tank filled with sea water and photographing small bubbles that were moved perpendicularly to the zone of focus while affixed to a fine glass fiber 32 μm in cross section held between two posts. Accurate positioning of the fiber relative to the camera was accomplished by moving the fiber support structure in relation to the lens using a set of vernier calipers secured to the housing. Photographs were taken as the bubbles that adhered to the glass fiber were moved intervals of 0.25 cm to and fro in front of the camera port. The resulting photographs were examined by microscope, and the limits of bubble focus then established according to the criterion of nonintersection of image dots. Finally, the use of high-contrast film for photographing bubbles eliminated the images of almost all other particles. Because solid particles in the sea absorb or scatter much of the light that strikes them, reflection of light from such particles produces film exposures that typically fall on the toe of the density versus exposure curve for the film. Such exposures are below the threshold necessary to produce a visible image. Bubbles, however, reflect incident light specularly and thus produce a strong image (ref. 33).

In field use, the camera system was suspended from a large float and allowed to drift freely. An initial 30-min delay in the timing circuit provided time for deployment of the apparatus before the first exposure, and permitted the movement of the ship downwind at least 0.5 km. Following the delay, photographs were made at 30-sec intervals until the entire roll of film was exposed. Following development, bubble images on the film were measured directly by microscope with the aid of an ocular micrometer (ref. 33).

With the above-described instrumentation and techniques, Johnson and Cooke carried out a detailed photographic analysis of microbubbles generated in coastal seas by breaking waves and general turbulence. As one might expect from the other studies of oceanic microbubble populations presented in previous sections of this chapter (see Sections 1.4.1, 1.4.2, and above), Johnson and Cooke (ref. 33) found that marine microbubble populations were always quite extensive, but varied considerably in total number depending upon the following factors: The microbubble diameter range examined (especially the minimum size), ocean location, water depth, wind speed, and wave conditions. For example, these authors first measured populations of bubbles at 1.5-m depth on March 10, 1977 outside St. Margaret's Bay, Nova Scotia, after the wind had been blowing for several hours at speeds of 8-10 m/sec. In the resulting population, all bubbles sampled were less than 250 μm in diameter, and nearly 75% of the total observed population was present as microbubbles in the diameter range of 50 to 120 μm. The concentration of bubbles greater than 34 μm in diameter under these conditions was found to be 2.7 x $10^4/m^3$ (ref. 33). In comparison, Johnson and Cooke point out that under nearly the same conditions of wind speed and water depth, Medwin (ref. 31) found in his early study, using a photographic technique but at a different ocean location (see above), more than 10^6 microbubbles per cubic meter in the diameter range of 30-100 μm.

Besides the importance of ocean location, Johnson and Cooke (ref. 33) provide data on the effect of water depth on microbubble concentration. Bubble populations were determined on April 3, 1977 in St. Margaret's Bay, Nova Scotia, at wind speeds of 11-13 m/sec. White capping was reported to be extensive, and the tops of waves were blown off during gusts. Under these conditions and at water depths of 0.7, 1.8, and 4 m, 60% or more of the total bubble population measured in each of the three cases fell within the (microbubble) diameter range of 50 to 120 μm, with the peak of the bubble distribution occurring consistently (and rather sharply) at about 100 μm diameter. The minority of bubbles with diameters over 120 μm decreased progressively with depth, as was expected due to the rise and loss of the larger bubbles. The actual concentration of bubbles greater than 34 μm in diameter, under these ocean conditions, was determined to be $4.8 \times 10^5/m^3$ at 0.7-m depth, $1.6 \times 10^5/m^3$ at 1.8-m depth, and $1.6 \times 10^4/m^3$ at 4-m depth.

In a subsequent study, Johnson and Cooke (ref. 42) hypothesize that these large and relatively persistent marine bubble populations primarily represent stabilized microbubbles, which are a common feature of the near surface region of the oceans. In support of their belief, the authors present a photographic record of the formation of stable spherical microbubbles and report their temporal and mechanical stability in samples of natural sea water. For these experiments, an observation cell with glass internal surfaces was constructed to allow viewing and photomicrography. Before an experiment began, all cell and container surfaces with which seawater made contact were thoroughly cleaned in strong detergent solution or hot chromic acid, rinsed repeatedly with deionized and filtered distilled water, and rinsed a minimum of three times with a sea water sample. The sea water used had been filtered through preoxidized glass fiber filters beforehand, then partially degassed, and finally equilibrated at 22°C with the atmosphere; this sequence ensured that equilibration was approached from undersaturation. Newly formed microbubbles with initial diameters between 40 and 100 μm were then introduced into the observation cell, which had been filled with the equilibrated seawater. When the microbubbles rose within the cell and to the underside of its viewing plate, they were examined with a photomicroscope capable of resolving bubbles and particles as small as 0.5 μm in diameter. The temperature remained constant (at 22°C) throughout each experiment (ref. 42).

Once in the observation cell, all microbubbles began to dissolve spontaneously because of their surface tension, and they diminished in size at an increasing rate as dissolution progressed. Johnson and Cooke observed that the dissolution rate became very rapid once the microbubbles reached about 10 μm in diameter, and those microbubbles that dissolved completely did so in less than 10 seconds. When a microbubble dissolved completely, a small transparent particle composed of the material originally present on the microbubble at the air/water interface always remained -- a familiar result following complete bubble dissolution (see Section 1.1.4). Many microbubbles did not dissolve completely, however. They stopped decreasing in size abruptly, sometimes becoming slightly aspherical, and remained as long-lived microbubbles apparently stabilized by films compressed during dissolution. These stabilized microbubbles were easily distinguishable from the transparent particles. The diameters of nearly 800 of these stabilized microbubbles were measured by Johnson and Cooke with an ocular micrometer and found to range from under 1 μm to 13.5 μm (ref. 42). (Presumably, if these authors had begun each experiment with newly formed bubbles much larger than the 40- to 100-μm diameter microbubbles they employed, the final diameters of the stabilized microbubbles produced and trapped in the sea water sample would have ranged considerably higher than only 13.5 μm.) Johnson and Cooke re-emphasize the common argument that the small sizes of these stable microbubbles theoretically mandates that they dissolve rapidly in response to their surface tension pressures, which are typically about 0.2 atm for a microbubble of 13.5 μm, and nearly 3.0 atm for a microbubble of 1 μm in diameter. These authors conclude that some mechanical or diffusion-limiting effect that resists further microbubble dissolution is activated during bubble contraction and the attendant compression of encapsulating surface films (ref. 42).

To further study the physical stability of these long-lived microbubbles, Johnson and Cooke also applied small negative and positive changes in pressure to the contents of the observation cell. When sufficient negative pressure was applied, the microbubbles expanded. These authors observed that expansion proceeded slowly at first, and then more rapidly. When the negative pressure was removed, some microbubbles contracted and dissolved completely, but most returned to nearly the same sizes and shapes that they had as stabilized microbubbles. In separate experiments, the application of pressure was found to result in rapid and complete dissolution of some of the (previously) stable microbubbles. For example, Johnson and Cooke report that, in one experiment, a pressure equivalent to 0.83 m of sea water reduced a population of film-stabilized microbubbles by 74%. Furthermore, there was size-dependent resistance to pressure apparent in the size distribution of the surviving microbubbles: bubbles 0.75 to 2.25 μm in diameter were reduced in number by slightly more than a factor of ⅓, while those greater than about 5 μm were reduced by a factor of more than ten (ref. 42). (It is useful to note that the above observations, concerning the dynamic behavior of film-stabilized microbubbles, mimic closely various predictions made by Bernd more than a decade before (ref. 16,17) on the basis of his earlier-described physical model for film-stabilized cavitation nuclei (see Section 1.3.1).)

As one might expect, out of the newly formed microbubbles put into the observation cell, the percentage that formed stabilized microbubbles changed little when the experiment was repeated in the same sea water sample, but varied between 2 and 93 percent in sea water samples taken over the period from 3 to 25 November 1980. Specifically, in samples taken on 3, 4, and 7 November by Johnson and Cooke (ref. 42), stable microbubbles were formed by 24, 25, and 93 percent of the microbubbles examined, respectively. However, in two succeeding experiments on 13 and 25 November, the percentage of stabilized microbubbles fell to only 2 percent. From these results, Johnson and Cooke postulated that this decrease was most likely due to a change in the concentration or character of the natural surfactants in the sea water, a consequence of increasingly stormy weather and the seasonal decline of biological activity in Labrador Current water over this period (ref. 42). (These circumstances strongly suggest a parallel between film-stabilized microbubble formation and the seasonal occurrence of natural organic particles (ref. 42,79; see also Section 1.1.4).)

In an additional experiment designed to test the persistence of film-stabilized microbubbles as a function of time, a population of stabilized microbubbles generated by Johnson and Cooke was maintained at 22°C in the observation cell and examined periodically for changes in size and number. After 4 hours there were no apparent changes; after 22 hours, while there was little reduction in number, these film-stabilized microbubbles generally were smaller, and those that previously were aspherical had become less so. Finally, 30 hours after formation, few visible microbubbles remained (ref. 42).

From all of their above-described studies, Johnson and Cooke conclude that since a significant proportion of the bubbles produced by breaking waves at sea are smaller than 200 μm in diameter (ref. 33,75), and because small bubbles dissolve in air-saturated sea water (when biological surfactant concentrations are low) as a result of surface tension alone, the number of film-stabilized marine microbubbles that are produced by breaking waves can be very large and show strong periodicity (ref. 42).

Chapter 2

EARLY WORK WITH AQUEOUS CARBOHYDRATE GELS

For more than a decade, bubble formation in gelatin had served as a helpful model for studying the etiology of decompression sickness in humans (ref. 34,35,135-137). From these and other studies, it has become clear that bubbles originate in water and possibly also in human tissues as long-lived gas microbubbles, which must be held intact by surface-active organic films (ref. 29; see also Chapters 1 and 8). In order to further study the physicochemical properties of the surfactant monolayer films surrounding gas microbubbles, a different gel method was developed, during 1975-76, which utilizes agarose as the gelling substance (ref. 138,139). Unlike gelatin, which is composed of a complex mixture of charged peptide chains (ref. 140), agarose is an uncharged, relatively inert, homogeneous polysaccharide (ref. 141) that is commercially available in highly purified form. Consequently, agarose gels make possible well-controlled, surface-chemical studies on microbubble stabilization and related bubble growth.

2.1 DEVELOPMENT OF THE AGAROSE GEL METHOD FOR MONITORING BUBBLE FORMATION

A frequent complaint with the gelatin method has been that different batches of the same commercial type gelatin powder would yield bubble counts, upon decompression of the resulting gels, that differed by as much as three orders of magnitude even though the exact same pressure schedule and pure gas (nitrogen) was used (D'Arrigo, unpublished data). Furthermore, even within the same batch of powder, it was found that when gelatin was freshly mixed prior to each test schedule, the variability of the results was intolerable. To improve the reproducibility, a single large batch of gelatin needed to be mixed and stored in a deep freeze (ref. 135). The preceding technical problems almost certainly (cf. below) derive from the fact that gelatin is composed of a complex mixture of charged peptide chains along with a variety of inorganic, ionic chemical contaminants (ref. 140).

In an attempt to avoid many of the above problems, a different gelling medium which was better defined chemically was sought. A literature search revealed that agarose is an uncharged, relatively inert, homogeneous polysaccharide (ref. 142-145) which forms reasonably transparent gels. The substance is also commercially available in several grades of purity, a fact which allowed at least partial insight into the effect of chemical impurities on bubble formation. Two brands of moderately pure agarose (i.e., from Sigma Chemical Co. and Miles Research Products) were tested first. A substantial improvement over gelatin in regard to the consistency and reproducibility of bubble counts was noted, but reproducibility was still not adequate when the agarose powder (from any given lot number) was newly dissolved prior to each test pressure schedule. Subsequently, a far purer grade of agarose (Bio-Rad) was located. (The specifications on this grade included two significant features, i.e., low sulfate ($< 0.3\%$, or expressed as sulfur, $< 0.1\%$) and ash ($< 0.5\%$) contents, that will be discussed below.) Accordingly, with this Bio-Rad agarose, good reproducibility of results was obtained uniformly using newly dissolved powder (1% w/w) from the same chemical lot number (i.e., 14672). Consequently, this specific chemical batch was used in all the initial experimental work on bubble formation in agarose gels (ref. 138,139).

The pressure vessel, glass counting chambers, and their plexiglass holder used in the experiments reviewed below are all similar to those described elsewhere (ref. 114,139). As viewed through the plexiglass wall of the pressure vessel, the inside dimensions of each rectangular counting chamber are 15 mm wide and 6 mm along the line of sight. (Preliminary preparation of the agarose solutions themselves was always carried out in glass containers, since plastic containers of various types were almost always found in these surface-chemical experiments to contribute problematic quantities of plasticizers to the aqueous solutions. A similar problem has been noted before by other investigators (e.g., ref. 146), but more often with regard to the use of organic solvents.)

Each counting chamber was filled with agarose solution to a

depth of 7 mm. After gelation, the agarose samples were exposed to a fixed pressure schedule (i.e., 0 psig (held for 22 hr at 21°C), then pressure increased at 1 psig/sec $\Rightarrow$ "saturation pressure" (held for 22 hr at 21°C), then rapid decompression (within 15 sec) $\Rightarrow$ 0 psig) using pure gases. In control experiments, the 22-hr saturation period was found to be more than adequate to completely saturate the aqueous gel medium (ref. 139); however, only bubbles formed between the depths of 1 and 4 mm in the gel were counted. This procedure avoided any extraneous effects on bubble number which might arise either from the observed deposition of precipitates at the bottom of some counting chambers after chemical modification of (i.e., electrolyte addition to) the sol prior to gelation or from the mere presence of a gas/gel interface at the top of the sample. Therefore, the total volume of gel examined in each sample amounted to 0.27 ml. In this specified volume, all bubbles formed appeared randomly distributed and none of them were in contact with the walls of the glass counting chambers. (The one exception to the above methodological description involved all the CAV-CON gel samples (lot no. 1005; see Table 2.5 below), which were supplied by the manufacturer in rectangular thermoplastic cuvettes with optical windows on two opposite sides. Each cuvette had a 10 x 10 mm interior bottom and contained 1.0 ml of agarose gel; furthermore, all bubbles (below a horizontal plane tangent to the bottom of the curved gas/gel interface) were counted since, despite provision of the same spectrum of electrolyte additions mentioned above, each of the specially purified CAV-CON agarose gel samples (0.2 % w/w) were without detectable turbidity or sediment. Finally, the above-mentioned 22-hr saturation period was found to be entirely adequate to saturate the approximately 10-mm deep CAV-CON gels.)

Experimental work was accordingly carried out (ref. 139), using this newly developed agarose gel method, to determine some of the physicochemical properties of the surfactants believed to stabilize the long-lived gas microbubbles in these aqueous gels. As a starting point, Fig. 2.1 summarizes the results obtained from a variety of rapid decompressions to atmospheric pressure after saturation of agarose samples with either N_2 (part A), CO_2 (part B), or He (part C). The ordinate plots the observed number of bubbles (mean of four trials) on a logarithmic scale. For the sake of visual

38

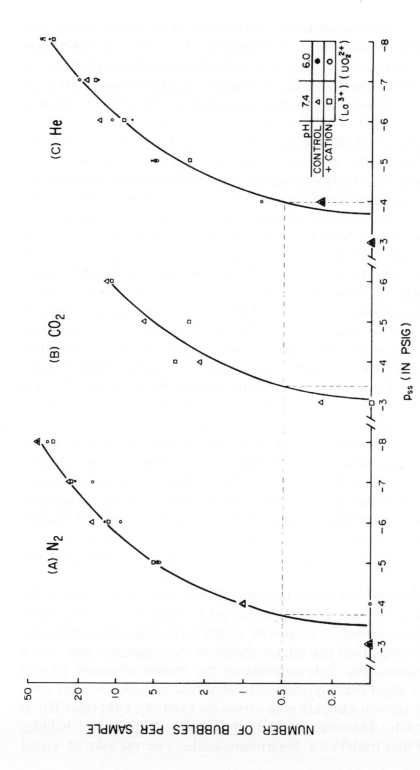

Fig. 2.1. Formation of bubbles in agarose gels as a result of rapid decompression to atmospheric pressure after saturation with different gases. (Taken from ref. 139.)

clarity, the corresponding standard errors of the means have been listed separately in Table 2.1 along with the values from individual trials. The abscissa plots the magnitude of the decompression, or supersaturation pressure, p_{ss}, in psig. For a given gas, the different symbols represent different chemical modifications of the agarose gel medium. It can be seen from the similarity of the curves for the three gases, representing the best visual fits to their respective points, that bubble <u>number</u> varied predominantly as a function of decompression magnitude (p_{ss}) and was virtually independent of the particular gas used. However, of the three gases tested, it was qualitatively observed that the CO_2 bubbles grew the fastest and to the largest size following any given decompression of the aqueous gels. In fact, with decompressions in excess of -6 psig, coalescence of CO_2 bubbles occurred so quickly that accurate bubble counts could not be obtained (ref. 139).

The actual threshold for bubble formation near atmospheric pressure in agarose gel is also represented in Fig. 2.1 for each gas. The approximate value is given by the intersection of the horizontal dotted line at the "0.5 bubble" level, i.e., 1 bubble in 50% of trials, with any of the curves. Inspection of the abscissas at the points of intersection indicates cavitation thresholds of between -3 and -4 psig ($\approx$ -0.25 atm) for all three gases. This range of cavitation thresholds was observed in additional experiments, shown in Fig. 2.2, to be the same for decompressions starting from atmospheric pressure or ambient pressures as high as 5 atm (ref. 49). The data in Fig. 2.2 is for CO_2, and very similar bubble numbers were also noted using either N_2 or He, or a 50%:50% mixture of CO_2 and N_2 (ref. 49).

2.2 RESULTS FROM DILUTE ELECTROLYTE ADDITIONS AND pH CHANGES IN AGAROSE GELS

Detailed inspection of Fig. 2.1 and Table 2.1 reveals that addition of 1 mM La^{3+} or 10 μm UO_2^{2+} to the agarose gel medium had only slight effects, if any, on the number of bubbles produced by a given decompression, regardless of the gas used (ref. 49,139). Similarly, lowering the pH of the gel medium (from 7.4) to 6.0 (Fig. 2.1) or even to 4.0 (ref. 147) had little or no effect on bubble formation.

UO_2^{2+}, at the concentration used in the above-mentioned

TABLE 2.1

Number of bubbles (four trials; mean $\pm$ S.E.M.) formed from rapid decompressions to atmospheric pressure of agarose samples (0.27 ml in volume) after saturation with either N_2, CO_2, or He. (Taken from ref. 139.)

Gas	p_{ss} (in psig)	pH 7.4		pH 6.0	
		control	1 mM La^{3+}	control	10 µM UO_2^{2+}
N_2....	-3	0	0	0	0
	-4	0,2,1,1 1.0$\pm$0.41	0,1,1,2 1.0$\pm$0.41	0,0,0,1 0.25$\pm$0.25	0
	-5	2,5,6,6 4.75$\pm$0.95	5,4,7,4 5.0$\pm$0.71	5,6,16,8 8.75$\pm$2.50	3,5,4,6 4.5$\pm$0.65
	-6	14,19,15,16 16.0$\pm$1.08	9,15,11,12 11.75$\pm$1.25	9,15,10,16 12.5$\pm$1.76	13,6,8,10 9.25$\pm$1.49
	-7	19,29,25,24 24.25$\pm$2.06	23,22,24,25 23.5$\pm$0.65	21,21,25,20 21.75$\pm$1.11	10,18,15,18 15.25$\pm$1.89
	-8	41,43,49,46 44.75$\pm$1.75	33,35,27,36 32.75$\pm$2.02	46,43,49,39 44.25$\pm$2.14	27,42,39,36 36.0$\pm$3.24
CO_2...	-3	1,0,0,0 0.25$\pm$0.25	0		
	-4	0,3,2,4, 2.25$\pm$0.85	1,4,4,5 3.5$\pm$0.87		
	-5	6,6,8,5 6.25$\pm$0.63	4,4,2,1 2.75$\pm$0.75		
	-6	14,11,13,12 12.5$\pm$0.65	12,11,9,13 11.25$\pm$0.85		
He....	-3	0	0	0	0
	-4	1,0,0,0 0.25$\pm$0.25	1,0,0,0 0.25$\pm$0.25	0,1,0,0 0.25$\pm$0.25	1,0,1,1 0.75$\pm$0.25
	-5	6,2,6,7 5.25$\pm$1.11	2,4,2,3 2.75$\pm$0.48	5,5,7,5 5.5$\pm$0.50	6,3,5,6 5.0$\pm$0.71
	-6	14,12,20,12 14.5$\pm$1.90	8,9,11,9 9.25$\pm$0.63	6,9,5,11 7.75$\pm$1.38	4,8,12,21 11.25$\pm$3.64
	-7	18,20,15,21 18.5$\pm$1.32	13,17,13,20 15.75$\pm$1.70	16,14,16,14 15.0$\pm$0.58	19,21,21,24 21.25$\pm$1.03
	-8	42,45,32,41 40.0$\pm$2.80	34,36,31,35 34.0$\pm$1.08	32,43,34,35 36.0$\pm$2.42	49,36,45,41 42.75$\pm$2.80

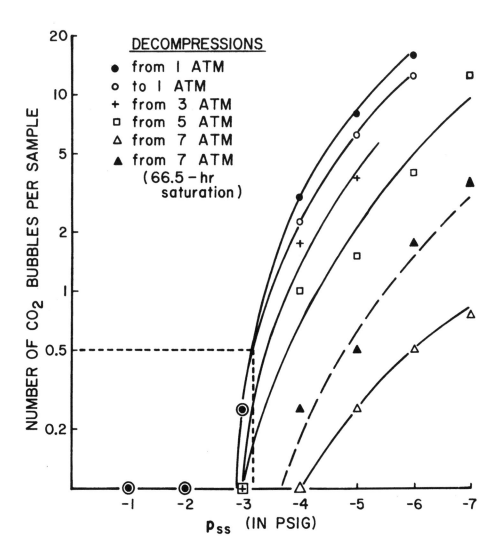

Fig. 2.2. Formation of CO_2 bubbles in agarose gels from rapid decompressions. The ordinate plots the observed number of bubbles (mean of 4 trials) on a logarithmic scale. The abscissa plots the magnitude of the decompression (or supersaturation pressure, p_{ss}, in psig). (Taken from ref. 49.)

experiments, is known to bind strongly and specifically to phosphate groups (ref. 148). If phospholipids represented a significant percentage of the surfactant molecules surrounding and stabilizing gas microbubbles, UO_2^{2+} would have bound to the polar complexes of these molecules and thereby likely have altered their surface-active properties (ref. 149-151); however, the stability of the gas microbubbles did not appear to be altered significantly since no appreciable change in bubble number occurred. More generally, if many of the surfactants in the films surrounding gas microbubbles possessed acidic groups of any kind, La^{3+} and/or H^+, in the concentrations used, would have very likely bound to these anionic groups (ref. 152); again, the bubble counts did not offer evidence that such binding had occurred to an appreciable extent (Table 2.1). In conclusion, these initial results collectively indicated that the surfactant monolayer surrounding a stable gas microbubble is comprised almost exclusively of nonionic surface-active molecules (ref. 139).

2.3 RESULTS FROM CONCENTRATED ELECTROLYTE ADDITIONS AND 1% PHENOL IN AGAROSE GELS

Follow-up experiments in this laboratory using other chemical batches of the same grade Bio-Rad agarose (lot numbers 15690 and 16320) had shown that although all results obtained followed exactly the same pattern as the above bubble counts (and hence were consistent with all the conclusions drawn from that published data), the precise bubble number obtained from any particular experiment can vary up to ten-fold depending upon which chemical batch of Bio-Rad agarose was employed. In order to proceed further with the chemical studies of the surfactant films surrounding gas microbubbles, a still purer grade of agarose was sought. Eventually an acceptable candidate, HGTP agarose (from Marine Colloids Div., FMC Corp.), was located. Upon testing, HGTP agarose gave highly reproducible bubble counts when newly dissolved powder (0.2% w/w) was used from the same chemical batch. More important, bubble counts (mean of 4 trials) for a given pressure schedule were found to be reproducible to within 10% across batches (lot numbers 1477 and 62867). In trying to explain this gratifying experimental result in chemical terms, it was noted from the manufacturer's specifications (ref. 145) that the HGTP agarose had both a lower sulfate content (0.13%) and a lower ash content (0.25%) than the Bio-Rad agarose (0.36% and 0.47%, respectively) used earlier. A technical report (ref. 145) by the manufacturer confirms the view that the low ash content indicates a lower content of inorganic salts, contaminants which are incompletely removed in processing, and also of cations associated with the residual anionic groups of the agarose. (Anionic residues such as sulfate and pyruvate are the primary contributors to this cation attraction (ref. 145).)

Using this highly purified HGTP agarose, a further evaluation was conducted of the above-mentioned tentative conclusion, from the initial phase of the agarose work (ref. 138,139), that the surfactant monolayers surrounding long-lived gas microbubbles do not contain a significant number of ionized chemical groups (see Section 2.2). If this conclusion was true, the suspected nonionic surfactants stabilizing the microbubbles should nevertheless be salted out from the aqueous gels if exposed to ion concentrations reaching into the decimolar range. Accordingly, this next agarose study (ref. 153) involved a systematic examination of the effects of a wide variety of salts, at a fixed concentration of 0.4 M, on bubble production within these high-grade agarose gels (following N_2 saturation at 85 psig).

All of the individual salt solutions used ranged in pH between 3 and 8; those solutions with a pH below 5 were brought to this H^+ concentration by neutralization with small amounts of NaOH. This was done since it had been found in control experiments that bubble production in agarose gels, in response to a fixed pressure schedule, did not vary significantly in the pH range of 5-8. At the concentration used (i.e., 0.4 M), the various salts tested were completely soluble at the pH's of their respective test solutions (ref. 152). (Equimolar concentrations of salts were used throughout these experiments so that the data obtained from this study would be directly comparable to the majority of the related salt data analyzed in the physicochemical literature (see below).) Finally, any effects of the different electrolytes on the solubility of nitrogen (as opposed to the solubility of nonionic surfactants) in the agarose gels were considered to be inconsequential as concerns these bubble experiments; it had been found in earlier studies using agarose gels (ref. 49,139) that, for all moderate-sized decompressions as used here, bubble number was virtually independent of the absolute amount per se of dissolved gas in the surrounding medium.

The results obtained from the outlined series of experiments supported the original (tentative) conclusion; it was found that the relative effectiveness of the different ions tested, in decreasing the degree of bubble formation in the aqueous gels (Table 2.2), exhibited many similarities with published data in the physico-chemical literature (ref. 154-179) for salting out of identified nonionic surfactants. Briefly, the pronounced and very similar anion sequences obtained (using sodium and potassium salts), combined with earlier data, suggested that the polar portions of these nonionic surfactants represent either weak bases and/or amide groups (ref. 153). This view was in accordance with the relative position of Mg^{2+} in the two cation sequences obtained (using chloride and nitrate salts), which indicated strong salting out in both cases, and hence rendered it quite unlikely that ether linkages contribute to the hydrophilicity of the nonionic surfactants stabilizing gas microbubbles (ref. 153).

In a subsequent study (ref. 180), three additional commercial preparations of ultrapure agarose were compared (Tables 2.3-2.5) for each of the 42 different electrolytes tested in the earlier investigation (ref. 153), and for one selected organic compound (Table 2.6), in order to better identify reproducible and significant trends in the salting-out data on natural, long-lived gas microbubbles. The data in Tables 2.3-2.5 demonstrate, as was also seen in the earlier study (Table 2.2), that the valence of the ions is not a factor of extreme and overriding importance as concerns the electrolyte effects on bubble production within the different commercial agarose gels. This finding confirms the belief (ref. 153) that the observed electrolyte effects are not the result of flocculation of charged colloids (ref. 128), which occurs through a nonspecific screening mechanism (ref. 128,163,164; cf. 181).

Other reasons to rule out a screening process as the explanation for these observed electrolyte effects (Tables 2.2-2.5) include the fact that ionic screening, which is based upon diffuse double-layer theory (ref. 182-186), would only occur if the net charge density on the microbubble monolayer-membrane surface is high (cf. ref. 181,187-189). Contrariwise, the data described in the preceding section (see Section 2.2) already indicated that the surfactant monolayer surrounding a stable gas microbubble is essentially comprised of nonionic molecules. Moreover, there are certain specific experimental predictions one could make if a pure screening process, rather than salting out, was the supposed basis

TABLE 2.2

Bubble production in Marine Colloids agarose gels (lot no. 62867; 0.2% w/w; 0.27 ml) with different salts (0.4 M). (Taken from ref. 153.)

Salt	No. of trials	Bubbles (mean±S.E.M.)	Salt	No. of trials	Bubbles (mean±S.E.M.)
Na_3 citrate	12	3.00 ± 0.48	K_3 citrate	12	4.00 ± 1.17
Na_2SO_4	16	5.38 ± 0.76	K_2SO_4	16	9.75 ± 1.15
Na_2 tartrate	16	5.94 ± 0.76	KF	16	12.00 ± 1.44
NaF	16	8.63 ± 0.85	K_2 tartrate	12	12.67 ± 1.25
Na acetate	12	16.08 ± 1.67	K acetate	16	15.63 ± 1.79
NaSCN	12	18.75 ± 1.92	KSCN	16	16.06 ± 1.07
NaBr	12	20.25 ± 0.96	KBr	16	18.25 ± 1.27
$NaNO_3$	16	20.44 ± 1.33	KI	16	20.19 ± 2.03
NaI	12	20.50 ± 1.22	KNO_3	20	20.65 ± 1.28
NaCl	16	23.56 ± 2.52	KCl	20	25.65 ± 1.24
(agarose control)	140	23.89 ± 0.49	(agarose control)	164	23.32 ± 0.43
$CaCl_2$	16	18.31 ± 1.24	$Mg(NO_3)_2$	16	17.19 ± 1.55
$MgCl_2$	16	21.25 ± 1.31	$Sr(NO_3)_2$	16	17.81 ± 1.27
LiCl	16	23.00 ± 1.23	$NaNO_3$	16	20.44 ± 1.33
NaCl	16	23.56 ± 2.52	KNO_3	20	20.65 ± 1.28
$CaCl_2$	12	24.08 ± 1.39	$LiNO_3$	16	20.88 ± 1.20
NH_4Cl	12	25.08 ± 1.23	$AgNO_3$	16	22.81 ± 1.84
KCl	20	25.65 ± 1.24	NH_4NO_3	12	23.17 ± 1.36
$BaCl_2$	12	32.25 ± 1.64	$Ca(NO_3)_2$	12	23.33 ± 1.53
$NiCl_2$	16	40.44 ± 2.32	$Cd(NO_3)_2$	16	26.69 ± 1.40
$SrCl_2$	16	49.69 ± 6.06	$Zn(NO_3)_2$	16	36.25 ± 2.41
$CoCl_2$	16	59.06 ± 13.8	$Ni(NO_3)_2$	16	67.25 ± 11.2
(agarose control)	168	25.17 ± 0.49	(agarose control)	172	24.69 ± 0.34

TABLE 2.3

Bubble production in Miles Lab. agarose gels (lot no. 1477; 0.2% w/w; 0.27 ml) with different salts (0.4 M). (Taken from ref. 180.)

Salt	No. of trials	Bubbles (mean$\pm$S.E.M.)	Salt	No. of trials	Bubbles (mean$\pm$S.E.M.)
Na_3 citrate	12	2.00 $\pm$ 0.51	K_3 citrate	12	3.08 $\pm$ 0.62
NaF	12	3.58 $\pm$ 0.51	K_2SO_4	12	7.33 $\pm$ 1.04
Na_2SO_4	12	4.50 $\pm$ 0.54	K_2 tartrate	12	8.83 $\pm$ 0.81
Na_2 tartrate	12	7.08 $\pm$ 0.58	KF	12	9.17 $\pm$ 0.60
Na acetate	12	8.42 $\pm$ 1.25	K acetate	12	10.00 $\pm$ 1.33
NaCl	12	13.08 $\pm$ 1.92	KSCN	12	16.67 $\pm$ 1.21
$NaNO_3$	12	17.83 $\pm$ 1.65	KNO_3	12	17.42 $\pm$ 1.07
NaSCN	12	19.08 $\pm$ 1.84	KBr	12	19.33 $\pm$ 1.20
NaI	16	19.69 $\pm$ 1.53	KI	12	21.75 $\pm$ 1.92
NaBr	16	21.75 $\pm$ 1.26	KCl	12	22.25 $\pm$ 1.32
(agarose control)	128	23.75 $\pm$ 0.51	(agarose control)	120	25.44 $\pm$ 0.43
NaCl	12	13.08 $\pm$ 1.92	$Zn(NO_3)_2$	12	16.33 $\pm$ 2.72
LiCl	12	15.33 $\pm$ 0.82	KNO_3	12	17.42 $\pm$ 1.07
$MgCl_2$	12	15.88 $\pm$ 1.33	$NaNO_3$	12	17.83 $\pm$ 1.65
NH_4Cl	12	15.75 $\pm$ 1.34	$Mg(NO_3)_2$	16	18.00 $\pm$ 1.14
$CdCl_2$	12	15.83 $\pm$ 0.95	$LiNO_3$	12	18.25 $\pm$ 1.67
$BaCl_2$	12	20.42 $\pm$ 1.20	NH_4NO_3	12	20.75 $\pm$ 1.59
$CoCl_2$	12	21.08 $\pm$ 1.61	$AgNO_3$	12	23.67 $\pm$ 2.10
$CaCl_2$	12	21.17 $\pm$ 1.08	$Ca(NO_3)_2$	12	25.00 $\pm$ 1.96
KCl	12	22.25 $\pm$ 1.32	$Cd(NO_3)_2$	12	28.67 $\pm$ 2.35
$NiCl_2$	12	29.17 $\pm$ 1.20	$Ni(NO_3)_2$	12	73.67 $\pm$ 3.29
$SrCl_2$	12	36.58 $\pm$ 2.03	$Sr(NO_3)_2$	12	107.17 $\pm$ 8.54
(agarose control)	132	27.24 $\pm$ 0.49	(agarose control)	136	26.84 $\pm$ 0.54

TABLE 2.4

Bubble production in FMC Corp. agarose gels (lot no. 60808; 0.2% w/w; 0.27 ml) with different salts (0.4 M). (Taken from ref. 180.)

Salt	No. of trials	Bubbles (mean±S.E.M.)	Salt	No. of trials	Bubbles (mean±S.E.M.)
Na_2 tartrate	12	9.67 ± 0.90	K_2SO_4	12	9.08 ± 0.82
Na_3 citrate	12	11.83 ± 1.18	K_3 citrate	12	11.17 ± 1.28
Na_2SO_4	12	14.00 ± 1.31	K_2 tartrate	12	11.92 ± 1.95
NaF	12	24.75 ± 2.48	KF	12	30.83 ± 2.53
Na acetate	12	27.50 ± 1.29	KNO_3	12	38.42 ± 2.13
NaCl	12	32.58 ± 4.20	K acetate	12	39.00 ± 1.44
NaSCN	12	39.92 ± 1.80	KCl	12	40.67 ± 2.79
$NaNO_3$	12	48.92 ± 2.33	KSCN	12	42.33 ± 2.27
NaI	12	49.00 ± 3.10	KI	12	47.58 ± 2.43
NaBr	12	52.75 ± 2.31	KBr	12	49.50 ± 2.39
(agarose control)	120	50.78 ± 0.71	(agarose control)	120	50.64 ± 0.62
NH_4Cl	12	30.08 ± 1.95	NH_4NO_3	12	35.92 ± 2.18
$MgCl_2$	12	31.33 ± 1.51	$Zn(NO_3)_2$	12	38.42 ± 2.07
NaCl	12	32.58 ± 4.20	KNO_3	12	38.42 ± 2.13
$BaCl_2$	12	35.58 ± 3.79	$LiNO_3$	12	44.08 ± 1.85
$CaCl_2$	12	37.25 ± 2.42	$Mg(NO_3)_2$	12	45.17 ± 2.08
KCl	12	40.67 ± 2.79	$Cd(NO_3)_2$	12	47.83 ± 3.43
$NiCl_2$	12	42.25 ± 2.42	$AgNO_3$	8	48.38 ± 2.62
LiCl	12	42.50 ± 2.71	$NaNO_3$	12	48.92 ± 2.33
$CoCl_2$	12	45.67 ± 2.85	$Ca(NO_3)_2$	12	52.00 ± 1.71
$SrCl_2$	12	51.42 ± 2.22	$Ni(NO_3)_2$	12	73.75 ± 3.86
$CdCl_2$	12	51.50 ± 2.47	$Sr(NO_3)_2$	12	88.92 ± 8.36
(agarose control)	132	50.66 ± 0.70	(agarose control)	128	50.55 ± 0.88

TABLE 2.5

Bubble production in CAV-CON agarose gels (lot no. 1005; 0.2% w/w in 0.95 ml samples) with different salts (0.4 M). (Taken from ref. 180.)

Salt	No. of trials	Bubbles (mean$\pm$S.E.M.)	Salt	No. of trials	Bubbles (mean$\pm$S.E.M.)
Na_3 citrate	12	2.17 $\pm$ 0.42	K_3 citrate	12	1.00 $\pm$ 0.39
Na_2 tartrate	12	2.92 $\pm$ 0.53	K_2 tartrate	12	2.58 $\pm$ 0.53
NaF	12	4.08 $\pm$ 2.14	K_2SO_4	12	4.83 $\pm$ 0.82
Na_2SO_4	12	5.08 $\pm$ 1.00	KF	12	6.42 $\pm$ 0.70
NaCl	12	14.50 $\pm$ 2.21	K acetate	12	17.50 $\pm$ 1.12
Na acetate	12	26.25 $\pm$ 2.97	KBr	12	21.50 $\pm$ 1.73
$NaNO_3$	12	31.17 $\pm$ 2.00	KCl	12	22.58 $\pm$ 2.25
NaBr	12	34.25 $\pm$ 3.21	KI	12	33.33 $\pm$ 2.14
NaI	12	36.42 $\pm$ 1.40	KNO_3	12	43.92 $\pm$ 2.45
NaSCN	12	43.33 $\pm$ 2.21	KSCN	12	48.33 $\pm$ 1.87
(agarose control)	120	43.63 $\pm$ 0.60	(agarose control)	120	46.71 $\pm$ 1.41
NH_4Cl	12	8.25 $\pm$ 1.41	NH_4NO_3	11	26.27 $\pm$ 4.10
NaCl	12	14.50 $\pm$ 2.21	$NaNO_3$	12	31.17 $\pm$ 2.00
KCl	12	22.58 $\pm$ 2.25	$Mg(NO_3)_2$	12	42.33 $\pm$ 3.86
$CaCl_2$	12	23.00 $\pm$ 1.26	KNO_3	12	43.92 $\pm$ 2.45
$BaCl_2$	12	24.83 $\pm$ 1.32	$LiNO_3$	10	47.40 $\pm$ 5.14
LiCl	12	26.83 $\pm$ 1.99	$Sr(NO_3)_2$	12	115.67 $\pm$ 13.1
$MgCl_2$	12	27.25 $\pm$ 2.78	$Zn(NO_3)_2$	12	135.08 $\pm$ 8.95
$NiCl_2$	12	49.33 $\pm$ 2.22	$Ca(NO_3)_2$	12	162.18 $\pm$ 27.0
$CdCl_2$	12	103.92 $\pm$ 10.8	$AgNO_3$	12	195.25 $\pm$ 16.5
$CoCl_2$	12	112.67 $\pm$ 28.8	$Cd(NO_3)_2$	12	212.25 $\pm$ 9.76
$SrCl_2$	12	124.58 $\pm$ 6.18	$Ni(NO_3)_2$	12	> 250
(agarose control)	120	44.92 $\pm$ 0.88	(agarose control)	132	44.26 $\pm$ 1.35

for the electrolyte effects summarized in Tables 2.2-2.5. These predictions derive from the fact that a pure screening process (described mathematically by Gouy-Chapman theory which is the simplest form of diffuse double-layer theory) is one in which the ions in solution do not approach closely enough to the oppositely charged groups on the membrane (i.e., monolayer) surface to give any importance to such factors as the radius, polarizability, or hydration energy of the ion (ref. 188). Therefore, the only factors determining the concentration of ions in the immediate vicinity of the monolayer surface would be the density of surface charges on the monolayer and the valence of the ions in solution. Accordingly, the primary expectation of a pure screening mechanism is that if an ion, e.g., monovalent cation, is substituted by an identical concentration of some other monovalent cation, the effect on the surface charges and potential ought to be the same since the binding energy of a particular ion species is no longer relevant (ref. 188). Hence, in the case of film-stabilized microbubbles, one would expect that identical extracellular concentrations of different cations with the same valence should have the same effect on bubble production. However, Tables 2.2-2.5 demonstrate that this was clearly not the case for monovalent cations, monovalent anions, divalent cations, or divalent anions with respect to bubble production in agarose gels. Instead, the anion and cation sequences obtained reflect salting out of the nonionic surfactants surrounding natural microbubbles.

It should also be emphasized that the pre-existing gas microbubbles examined in these agarose-gel studies are not associated with, and should not be confused with, the separate phenomenon of "spontaneous nucleation" (i.e., de novo formation or homogeneous nucleation) of bubbles (ref. 190-193). For de novo bubble formation to occur (as opposed to growth of pre-existing gas microbubbles), decompressions at least one order of magnitude greater (ref. 29,190-193) than those employed in these studies would be required. Consequently, the physicochemical basis of the electrolyte effects reported here is not explicable in the terms of classical nucleation treatments which do not consider the influence of surfactants. (For example, the progressive increase in bubble counts with each anion and cation sequence given in Tables 2.2-2.5 does not reflect an enhanced nucleation rate (often symbolized as I); any conceivable effect of the electrolytes upon either the frequency factor (z) or the positive free energy of formation of a bubble nucleus ($\triangle G_{max}$) appearing in the nucleation rate equation (ref. 194) [$I \approx Z \exp(-\triangle G_{max} / kt)$] would simply not be discernible with the relatively small decompressions employed in these studies.) Instead, the electrolyte effects on bubble formation described in these studies are quite consistent with the belief that the two anion and two cation sequences obtained reflect a decreasing tendency of the respective ions toward salting out (i.e., decreasing the solubility) of predominantly nonionic surfactants stabilizing the gas microbubbles present in agarose gels (ref. 153,180). Moreover, the precise pattern of these electrolyte (i.e., salting-out) effects on bubble formation can yield specific clues as to the structural characteristics of the nonionic surfactants stabilizing gas microbubbles present in agarose gels (see below).

2.4 DETAILED COMPARISON WITH PUBLISHED DATA IN THE PHYSICOCHEMICAL LITERATURE FOR SALTING OUT OF IDENTIFIED NONIONIC SURFACTANTS

Inspection of Tables 2.2-2.5 reveals that the anion sequences obtained with the different commercial agarose gels were more pronounced and far more similar, using sodium and potassium salts, than were the corresponding cation sequences obtained with chloride and nitrate salts. This result suggests (ref. 153) that the polar portions of these nonionic surfactants represent either weak bases and/or amide groups. Three additional results, however, weigh in favor of the amide group in particular.

First, the anion sequences obtained, along with the quantitative data presented in Tables 2.2-2.5, clearly suggest that polyvalent anions (e.g., citrate^{3-}, tartrate^{2-}, and SO$_4^{2-}$) have a moderately powerful salting-out effect on the nonionic surfactants stabilizing gas microbubbles. This finding is noteworthy since a variety of amides examined earlier by different investigators (ref. 195) showed similarly strong (yet not extreme) salting out by all polyvalent anions tested, and amide-containing nonelectrolytes are known to often possess surface-active properties (ref. 196,197).

Secondly, it can be seen from Table 2.6 that phenol (1% v/v) reduced bubble production, as compared with distilled/deionized water controls, by over 50% in all three of the ultrapure agarose

TABLE 2.6

Effect of phenol (1% v/v) on bubble production in different commercial agarose gels. (Taken from ref. 180.)

Agarose gel	Solution	No. of trials	Bubbles (mean±S.E.M.)
Miles Lab. (lot no. 1477; 0.2% w/w; 0.27 ml)	Distilled water control	12	22.25 ± 1.35
	Phenol (1% v/v in H_2O)	12	7.58 ± 1.20
FMC Corp. (lot no. 60808; 0.2% w/w; 0.27 ml)	Distilled water control	12	52.75 ± 3.24
	Phenol (1% v/v in H_2O)	12	18.83 ± 1.46
CAV-CON (lot no. 1005; 0.2% w/w; 0.95 ml)	Distilled water control	12	38.17 ± 6.65
	Phenol (1% v/v in H_2O)	12	18.50 ± 4.18

preparations available for this experimental study. The import of this observation derives from the fact that phenol, in aqueous solution at room temperature, is known to bind strongly to a variety of amide-containing compounds, including naturally occurring fibers (e.g., wool and horse hair) (ref. 198). Binding of phenol may be via the formation of hydrogen-bond complexes with alkylamide groups in these protein fibers in water (ref. 198). Refractive index measurements on aqueous solutions of simple amides (e.g., acetamide, dimethylformamide, and diacetamidomethane) and phenol have in fact suggested the existence of a 1:1 complex of the two solutes (ref. 199). The complex formation, however, also can very well involve an aromatic-amide interaction which may be ascribed to an interaction of the polarizable π electrons of phenol's aromatic ring with the amide or, possibly, to the formation of a weak molecular complex between the amide and unoccupied orbitals of the aromatic system (ref. 200). Regardless of the precise mechanism(s) of phenol binding, such complex formation with the nonionic surfactants stabilizing gas microbubbles can be expected to destabilize the presumed surfactant monolayer structure (ref. 29,139) surrounding these long-lived microbubbles by altering the hydrophilic/lipophilic balance (HLB) (ref. 150,201-206) of individual surfactant molecules at the gas/aqueous gel interface. The result might well be a sizable effect of phenol on bubble production in these aqueous gels as found in this study (Table 2.6).

The third finding which suggests that amide groups, and not weak bases, represent the polar portions of the nonionic surfactants stabilizing gas microbubbles concerns the observed effects of certain cations on bubble production in agarose gels. Although there was considerable variability among the cation sequences (using chloride and nitrate salts) obtained for any one of the agarose gel types tested (Tables 2.2-2.5), this variability was least with the specially purified CAV-CON agarose gel (Table 2.5) and sufficiently low to allow some useful information to be extracted. In particular, it was noted upon testing with the chloride and nitrate salts that this gel type yielded the same, and hence presumably more reliable, bubble-reduction (or salting-out) sequence for three important cations, namely, $Na^+ > K^+ > Li^+$ (Table 2.5). The importance of this particular finding stems from the following observation of Long and McDevit (ref. 155) described in their classic analysis of the relevant early literature concerning salt effects on nonelectrolyte solutes: "When the specific effects of salts on polar nonelectrolytes [e.g., nonionic surfactants] are considered more closely, it is found that, although in a gross sense the salt order is similar to that for nonpolar molecules, there are significant variations in detail. Furthermore, these variations fall into two distinct groups. For one, which includes nonelectrolytes that have a definite basic character, there is an increased sensitivity to changes in the anion of the salt and also a considerable shift towards increased salting in by lithium salts. In contrast, for the group which includes the weak acids, the salting out by lithium salts is increased, relatively, and is usually larger than for sodium salts. A relatively greater sensitivity for changes in the cation is observed for this group. Finally, there is an intermediate group where the salt order [i.e., $Na^+ > K^+ > Li^+$ regarding salting out] is similar to that for nonpolar electrolytes" (ref. 155). This salting-out sequence identified with intermediate polar nonelectrolytes is identical to the Na-K-Li sequence obtained with both chloride and nitrate salts using the specially purified CAV-CON agarose gel. Consequently, this finding argues against the notion that the polar portions of the nonionic surfactants stabilizing long-lived gas microbubbles represent weak bases. Instead, amide groups, which are essentially neutral but large dipoles, within the nonionic surfactants are much more likely to give rise to the $Na^+ > K^+ > Li^+$ bubble-reduction (or salting-out) sequence repeatedly observed using the CAV-CON gel.

2.5   CONCLUDING REMARKS

In view of the three separate findings described in Section 2.4, it is concluded that the polar portions of a significant fraction of the nonionic surfactants stabilizing long-lived gas microbubbles represent primarily amide groups. This view is also consistent with the relative position of Mg^{2+} in all the cation sequences. Despite much variability among the cation sequences obtained with the different commercial agarose gels, the relative position of Mg^{2+} repeatedly indicated appreciable (and often strong) salting out in each case (Tables 2.2-2.5). This last result provides some additional information about the structure of the nonionic surfactants in agarose gels. The magnesium cation is known to readily form an oxonium compound with the lone-pair electrons of ether oxygen atoms of (synthetic) nonionic polyoxyethylated surfactants to cause salting in (ref. 175,179). Therefore, in view of the consistent salting-out effect displayed by Mg^{2+} in agarose gels (Tables 2.2-2.5), one must conclude that ether linkages in all likelihood do not contribute to the hydrophilicity of the nonionic surfactants stabilizing microbubbles.

As ether linkages commonly occur in a wide variety of the synthetic nonionic surfactants (ref. 163,164,204,207-209) present in trace amounts in many different water sources throughout our environment (ref. 15,16,210-213), it appears that such artificially produced chemical contaminants do not contribute measurably to the stability of gas microbubbles in a variety of aqueous (agarose) gels and, perhaps, aqueous media in general. The nonionic or hydrophobic surfactants stabilizing long-lived gas microbubbles are probably mostly, if not all, of natural origin.

Chapter 3

COMPARISON OF AQUEOUS SOIL EXTRACTS WITH CAR-
BOHYDRATE GELS

Since the experimental work with agarose gels described in
Chapter 2 clearly indicated that the polar portions of some of the
surfactants stabilizing microbubbles represent primarily amide
groups (i.e., peptide linkages and/or acid amide groups), a
reasonable hypothesis was that some of the surfactants were
proteinaceous in nature. In such a case, their considerable hydrophobicity might well be explained in terms of hydrophobic amino
acid residues, e.g., aromatic amino acids, comprising a large
(and/or significant) structural portion of these surfactants which
surround microbubbles. To test this hypothesis, three different
types of protein-specific chemical tests were performed on long-
lived microbubbles derived from: 1) aqueous solutions of agarose
powder, and 2) filtered aqueous extracts of Hawaiian forest soil.

Forest soil was, in fact, found to be an unusually rich source
of stable gas microbubbles. This particular natural source was
arrived at indirectly on the basis of the earlier-mentioned reports
(see Section 1.4) that in the open sea, which is known to contain
dissolved (ref. 89,214-220) and nonliving particulate (ref. 41,79,
221-224) organic matter and specifically surface-active substances
(ref. 38,41,43-45,86,93), an extensive microbubble population has
been detected by various workers (ref. 30-33). Consequently, it
appeared likely that the naturally occurring, largely hydrophobic
surfactants stabilizing long-lived gas microbubbles in agarose gels
(ref. 180) were present not only in marine algae (from which the
gels were derived (ref. 142-145)), but probably also in seawater
and other environmental sources as well. Accordingly, a systematic screening of a wide variety of natural substances in this
laboratory revealed that (filtered) aqueous extracts of ordinary
forest soil contained unusually high concentrations of gas microbubbles. This finding was reassuring and consistent with the fact

that humic substances, similar (although not identical) to those occurring in soils, have been found to be widespread in sea water by various investigators (ref. 223-228). An ongoing experimental program was thereafter expanded to further characterize and compare chemically the surfactants stabilizing long-lived micro-bubbles in both agarose gels and forest soil extracts (see below).

3.1 FUNCTIONAL MICROBUBBLE RESIDUES IN SOIL AND AGAROSE POWDER

The observed presence of surfactant-stabilized gas "micro-bubbles" in damp (i.e., undissolved) soil, and even dry agarose powder, becomes understandable in light of various related studies (see Section 1.1.4) in the literature. For example, it was described in Chapter 1 that Johnson and Cooke have shown that microbubble dissolution in carefully filtered sea water always results in the formation of an "organic particle" (ref. 41,42,74). Furthermore, these authors also cite the experiments of Liebermann (ref. 84) on the rate of bubble dissolution in distilled water, who in a similar fashion found that a residue always remains at the site where a bubble dissolves. Specifically, Liebermann (ref. 84) states "In spite of precautions to preclude the introduction of contaminants, *a permanent insoluble deposit remained after bubble collapse in every trial.* Some (not all) of these collapse sites were demonstrated to be capable of regenerating bubbles exactly in the manner described previously" (i.e., regarding separate experiments involving chalk-dust contaminated water). These interesting bubble regeneration observations were reported as follows: "A small air bubble was introduced into the [water] droplet by means of a micropipette. The bubble was observed microscopically using a 1.9-mm oil immersion objective. The bubble was seen to dimin-ish and then vanish from sight; but remaining on the underside of the glass at the site of the bubble collapse was a microscopic [insoluble deposit]. ...The next stage in the experiment demonstrates the likely existence of a persistent submicroscopic bubble: the water droplet containing the insoluble nucleus ... was then subjected to a reduced pressure of approximately ¼ atmos. Almost immediately a visible bubble appeared. The bubble rapidly grew in size. ...Clearly the increase in size resulted from the water droplet becoming oversaturated [with air] at the reduced pressure.

...Once formed, a bubble 'nucleus' is exceedingly stable. ...There appears to be no time limit beyond which it is not possible to regenerate a bubble from a hydrophobic collapse site. In these experiments as much as 48 hours were permitted to elapse between bubble collapse and regeneration" (ref. 84; cf. synthetic microbubble disappearance and formation in Section 10.4).

Very similar observations on the rates of growth and collapse of air bubbles in distilled degassed water were later made by Manley (ref. 8), who also estimated the diffusion coefficient of air through the "organic skin" surrounding small bubbles. He states that "the diffusion rate of gas out of bubbles in undersaturated water is reduced to approximately 10% of the normal rate, [just before] the bubble disappears into solution. After the bubble has disappeared, it has been found that a deposit of impurity occurs underneath the Perspex plate and this deposit is known not to be present before the bubble was formed. ... A larger number of similar bubbles has been studied and it has generally been found that the volume of the deposit was approximately constant" (ref. 8). Furthermore, such newly formed, microscopic deposits appear to contain a minuscule air pocket since it was also found "that a long exposure of the water to a vacuum was sufficient to reduce the volume of the impurity considerably" (ref. 8).

Moreover, the existence of air-filled organic shells even in an apparently anhydrous state has been reported in a study by Garrett (ref. 229,230) of the retardation of water drop evaporation by monomolecular surface films. He "noted that for water drops coated with certain rigid monolayers, evaporation appeared to cease at diameters considerably greater than that calculated for the residue of organic material. This event occurred at about the time that the drop surface manifested a solid, wrinkled appearance resembling a plastic bag. The water had then *evaporated completely, leaving behind a vacant organic shell* [emphasis supplied] whose diameter was much greater than that of a condensed sphere of the same quantity of organic material" (ref. 230).

3.2 ADAPTATION OF (FILTERED) AQUEOUS SOIL EXTRACTS FOR USE WITH THE AGAROSE GEL METHOD

The pressure vessel, glass counting chambers, and their plexiglass holder used in these experiments were all similar to those described above (see Section 2.1). The actual saturation pressure used in all experiments employing agarose gels not containing soil extract was 85 psig. With this saturation pressure, it was found in control experiments that essentially all of the bubbles formed in the gels originated from the agarose powder itself and not the distilled water used to make the gels, i.e., bubble number was found to be directly proportional to agarose concentration (in the range 0.2%-0.8% w/w. In additional control experiments it was determined that, with the particular high-purity agarose powder used (see below) in the present experiments at a concentration of 0.2% or 0.3% w/w, no bubbles formed in the 0.27 ml agarose gels when saturation pressures below 40 psig were used. Consequently, only saturation pressures of 30 or 40 psig were used in all experiments employing agarose gels made with filtered (see below) aqueous extracts of forest soil (as opposed to distilled water). Owing to the unusually high concentrations of gas micro-bubbles occurring in the aqueous soil extracts, the 0.27 ml agarose gels formed from such extracts were found to contain enough large (and hence lower-cavitation-threshold) microbubbles, out of a probable spectrum of many sizes, to consistently yield 20-40 bubbles upon decompression from saturation pressures of 30 or 40 psig. Accordingly, chemical modification of the bubble production elicited by these low saturation pressures would then provide structural information about those long-lived microbubbles originating specifically from forest soil (as opposed to the agarose powder) (ref. 231).

The samples of aqueous soil extract, utilized in each of the three protein-specific tests, were all prepared in the same manner. Approximately 500 g of (damp) surface soil, obtained from a forested area on windward Oahu, was added to 2 liters of distilled water and refrigerated. The mixture was gently swirled at intervals for one day and then all undissolved material allowed to settle during the second day. Thereafter, the supernatant was carefully removed and filtered with suction through a precleaned Gelman Acropor 3 μm (pore diameter) membrane filter used in conjunction with a Gelman thick glass-fiber prefilter. Aliquots (8 ml each) were then pipetted into separate glass vials and stored frozen for later use over the following 3 months in the soil extract experiments. When thawed for an experiment, none of the aliquots of soil extract contained any sediment and, when used with different agarose concentrations to form gels in control measurements, yielded an essentially constant number of bubbles for a given saturation pressure ≤ 40 psig. The reproducibility of such control measurements, observed with each of several different samples of Hawaiian forest soil collected, provided further support for the belief that surfactant-stabilized gas microbubbles exist in high concentrations in such aqueous soil extracts *since all solid particulate debris larger than 3 μm had been removed previously* by the filtration process (ref. 231).

3.3 NINHYDRIN EFFECT ON BUBBLE FORMATION IN COMMERCIAL AGAROSE AND AQUEOUS SOIL EXTRACTS

The data presented in Table 3.1 demonstrate that 0.3% (w/v) ninhydrin has a marked effect at pH 5.0 on bubble production in agarose gels containing distilled water. This concentration of ninhydrin reduced bubble production at pH 5.0 to only one third of that observed in the control buffered at the same pH, whereas no significant effect at pH 8.0 was observed (Table 3.1). This pH dependence of the chemical effect is important for the reason that ninhydrin is known to react appreciably with α-amino acids, peptides, and proteins in various buffered aqueous solutions ranging in pH from 5.0 to a maximum of 7.0 (ref. 232,233) and specifically in aqueous solutions containing citrate buffer (as used in this study) at pH 5.0 (ref. 233). While ninhydrin is also known to react with many amino compounds and ammonia (ref. 233), such interfering

TABLE 3.1

Effect of 0.3% (w/v) ninhydrin on bubble production in buffered agarose gels (0.3% w/w, 0.27 ml) containing distilled water. (Taken from ref. 231.)

Gel treatment	No. of trials	Number of bubbles (mean $\pm$ S.E.M.)
pH 5.0 (0.1 M Na$_3$citrate) buffer	16	35.69 $\pm$ 2.67
pH 5.0 buffer + 0.3% ninhydrin	16	11.63 $\pm$ 1.09
pH 8.0 (0.1 M Na$_3$citrate) buffer	16	23.19 $\pm$ 2.58
pH 8.0 buffer + 0.3% ninhydrin	16	20.50 $\pm$ 2.83

60

substances are essentially absent from the highly purified SeaKem HGTP Agarose powder (ref. 145,234-240) that was used, which contains a very slight contamination from only anionic residues (i.e., sulfate and pyruvate) (ref. 142-145). It therefore appears clear that some of the surfactants stabilizing long-lived gas microbubbles in agarose powder are proteinaceous; this finding agrees well with the previous expeimental results from this laboratory (see Chapter 2) indicating that the polar portions of these naturally occurring, largely hydrophobic surfactants represent primarily amide groups (i.e., peptide linkages and/or acid amide groups).

In addition, ninhydrin was observed to have a similarly strong, pH-dependent effect on bubble production in agarose gels containing aqueous soil extract (Table 3.2). In this case, 0.3% (w/v) ninhydrin at pH 5.0 reduced bubble production, as compared with the control buffered at the same pH, by approximately 35% , whereas only a slight (if any) effect occurred at pH 8.0 (Table 3.2). Hence, many of the surfactants stabilizing long-lived microbubbles in aqueous soil extract also appear to be proteinaceous in nature. This finding is consistent with the fact that approximately one quarter to one half of the (nitrogenous) organic matter in soils is proteinaceous material (ref. 241-264) which contains a wide variety of α-amino acids that have been successfully identified in both Hawaiian soils (ref. 241,245-247,264) and soils from other locations (ref. 242-244,248-263). Most of this proteinaceous material appears to be absorbed (via hydrogen bonding (ref. 265-267))

TABLE 3.2

Effect of 0.3% (w/v) ninhydrin on bubble production in buffered agarose gels (0.3% w/w, 0.27 ml) containing aqueous soil extract. (Taken from ref. 231.)

Gel treatment	No. of trials	Number of bubbles (mean $\pm$ S.E.M.)
pH 5.0 (0.1 M Na$_3$citrate) buffer	16	32.94 $\pm$ 2.01
pH 5.0 buffer + 0.3% ninhydrin	16	21.63 $\pm$ 1.88
pH 8.0 (0.1 M Na$_3$citrate) buffer	16	36.56 $\pm$ 1.41
pH 8.0 buffer + 0.3% ninhydrin	16	33.69 $\pm$ 2.10

onto humic-acid polymers (ref. 252,253,258-260,265-269) thereby forming complexes resistant to decomposition by biological agencies (ref. 252,253,265-267,269). Furthermore, a humoprotein complex has been successfully isolated by Simonart et al. (ref. 265) from podzolic, meadow, and brown forest soils and the humoprotein content was observed to be highest in the forest soil. The proteinaceous fraction itself was subsequently isolated and its maximum absorption lay in the spectral region 260-280 nm (ref. 265). These last two findings regarding brown forest soil and the absorption maximum of the protein fraction suggested that the aforementioned hydrophobicity of the evidently proteinaceous surfactants participating in the stabilization of long-lived gas microbubbles, present in both Hawaiian forest soil and agarose powder, might well be explained as follows: hydrophobic amino acid residues, particularly aromatic amino acids which absorb maximally in the spectral range 275-280 nm (ref. 270), comprise a large and/or significant structural portion of the surface-active proteinaceous compounds which surround microbubbles.

3.4   PHOTOCHEMICAL EXPERIMENTS USING METHYL-ENE BLUE

To test the above-mentioned (aromatic-amino-acid) hypothesis, a series of photochemical experiments was conducted which involved the use of a common dye, methylene blue. This dye is known to be an effective sensitizer of the photooxidation of some amino acids; the specific nature of this reaction allows the degradation of certain amino acids in proteins (ref. 271). Much previous work by other investigators (ref. 271-273) has clearly demonstrated that the aromatic amino acids are particularly susceptible to photooxidative destruction in the presence of methylene blue. It was therefore quite informative to find, in this experimental study (Tables 3.3, 3.5 and 3.6), that intense (2 h) irradiation with visible light did, in fact, lead to a marked reduction in the bubble-production capability of (transparent) agarose gels containing distilled water to which a low concentration (100 μM) of methylene blue had been added. Specifically, at pH 5.0 (Table 3.3), 4.5 (Table 3.5), and 4.0 (Table 3.6), the dye along with visible-light irradiation resulted in bubble reductions (as compared to control gels having the same pH and irradiation treatment, but no dye) of

TABLE 3.3

Effect of 100 µM methylene blue dye on bubble production in agarose gels (0.2% w/w, 0.27 ml) containing distilled water and pH 5.0 (1 mM NaHCO$_3$) buffer. (Taken from ref. 231.)

Gel treatment	No. of trials	Number of bubbles (mean ± S.E.M.)
Nonirradiated	20	28.25 ± 1.42
100 µM dye; nonirradiated	20	23.05 ± 0.99
2-h (intense visible-light) irradiation	20	25.15 ± 1.19
100 µM dye + 2-h irradiation	20	13.95 ± 0.99

45%, 42%, and 42%, respectively, whereas the dye without an accompanying irradiation treatment resulted in bubble reductions (as compared to dye-free, nonirradiated controls) of only 18%, 26%, and 21%, respectively. The corresponding bubble-reduction percentages in agarose gels containing aqueous soil extract at pH 5.0 (Table 3.4) were 27% (irradiated) and 15% (nonirradiated). (In this case, the smaller percentages observed probably resulted from a reduction of the free methylene blue concentration due to binding (ref. 274) of the dye to humic-acid polymers.) It should be emphasized that the approximately twofold increase of the bubble-reduction percentage in both types of gels, and at each pH, when the visible-light irradiation treatment accompanied the use of

TABLE 3.4

Effect of 100 µM methylene blue dye on bubble production in agarose gels (0.2% w/w, 0.27 ml) containing aqueous soil extract and pH 5.0 (10 mM NaHCO$_3$) buffer. (Taken from ref. 231.)

	No. of trials	Number of bubbles (mean ± S.E.M.)
Nonirradiated	16	28.00 ± 1.37
100 µM dye; nonirradiated	16	23.81 ± 1.98
2-h (intense visible-light) irradiation	16	28.31 ± 1.76
100 µM dye + 2-h irradiation	16	20.69 ± 1.83

TABLE 3.5

Effect of 100 µM methylene blue dye on bubble production in agarose gels (0.2% w/w, 0.27 ml) containing distilled water and pH 4.5 (1 mM NaHCO$_3$) buffer. (Taken from ref. 231.)

Gel treatment	No. of trials	Number of bubbles (mean $\pm$ S.E.M.)
Nonirradiated	16	21.00 $\pm$ 1.41
100 µM dye; nonirradiated	16	15.44 $\pm$ 0.81
2-h (intense visible-light) irradiation	16	18.94 $\pm$ 1.20
100 µM dye + 2-h irradiation	16	11.00 $\pm$ 1.02

methylene blue strongly argues that a genuine photooxidative destruction of proteinaceous material occurred in the agarose gels. Accordingly, the reduction in bubble production almost certainly stems from destruction of gas microbubbles, i.e., from methylene blue-sensitized photooxidation of amino acid residues within the proteinaceous surfactants surrounding microbubbles. It is useful to point out that Knowles and Gurnani have reported that in acidic aqueous solution, methylene blue causes extensive photooxidation of the aromatic amino acid tryptophan (ref. 271). While other aromatic amino acids also displayed an appreciable sensitivity to the photochemical action of methylene blue, it was not nearly as strong as the photosensitivity of tryptophan (ref. 271), and this

TABLE 3.6

Effect of 100 µM methylene blue dye on bubble production in agarose gels (0.2% w/w, 0.27 ml) containing distilled water and pH 4.0 (1 mM NaHCO$_3$) buffer. (Taken from ref. 231.)

Gel treatment	No. of trials	Number of bubbles (mean $\pm$ S.E.M.)
Nonirradiated	16	17.94 $\pm$ 1.27
100 µM dye; nonirradiated	16	14.25 $\pm$ 1.18
2-h (intense visible-light) irradiation	16	20.00 $\pm$ 2.04
100 µM dye + 2-h irradiation	16	11.56 $\pm$ 1.00

same finding has been reported by different investigators working at other pH's (ref. 275-277). In view of all the above findings, it appeared quite probable that a significant fraction of the naturally occurring surfactants which surround and stabilize long-lived microbubbles are proteinaceous compounds which contain (and whose surface activity depend upon) aromatic amino acid residues that include tryptophan.

3.5   2-HYDROXY-5-NITROBENZYL   BROMIDE   EXPERI-MENTS

To evaluate the above-mentioned, more specific hypothesis concerning tryptophan content, additional experiments were carried out using the tryptophan-specific reagent 2-hydroxy-5-nitrobenzyl bromide.   This reagent is known to be highly selective for tryptophan, in proteins and peptides, at acidic pH values (ref. 278-282).   While it has been reported that 2-hydroxy-5-nitrobenzyl bromide reacts with sulfhydryl groups of proteins at higher pH's (ref. 278,280), this reaction is so slight that the reagent can safely be used to modify tryptophan residues in proteins and peptides (ref. 282).   It was therefore further assuring to find, in the present experimental study (Table 3.7), that 1.0 M 2-hydroxy-5-nitrobenzyl bromide did, in fact, have a striking effect at pH 5.0 on bubble production in agarose gels containing distilled water.   Specifically, the reagent at pH 5.0 reduced bubble production, as compared with the control buffered at the same pH, by over 95%, whereas at pH 8.0 the bubble reduction was far less (i.e., 60%).   (The much

TABLE 3.7

Effect of 1.0 M 2-hydroxy-5-nitrobenzyl bromide (HNBB) on bubble production in buffered agarose gels (0.3% w/w, 0.27 ml) containing distilled water. (Taken from ref. 231.)

Gel treatment	No. of trials	Number of bubbles (mean $\pm$ S.E.M.)
pH 5.0 (0.1 M Na$_3$citrate) buffer	16	9.50 $\pm$ 1.14
pH 5.0 buffer + 1.0 M HNBB	16	0.31 $\pm$ 0.15
pH 8.0 (0.1 M Na$_3$citrate) buffer	16	11.69 $\pm$ 1.47
pH 8.0 buffer + 1.0 M HNBB	16	4.00 $\pm$ 0.77

smaller, but still quite noticeable, effect of the reagent at pH 8.0 is very likely not due to a tryptophan-specific chemical reaction (ref. 278-282) but rather physical absorption (ref. 282) of 2-hydroxy-5-nitrobenzyl bromide on the surface of the proteinaceous surfactants surrounding microbubbles.) The observed pH dependence of the 95% bubble reduction achieved with 2-hydroxy-5-nitrobenzyl bromide renders it virtually certain that the proteinaceous surfactants participating in the stabilization of long-lived gas microbubbles in agarose powder contain tryptophan residues. These residues appear to be essential for the normal surface-active behavior of these proteinaceous compounds.

In addition, 2-hydroxy-5-nitrobenzyl bromide was observed to have a similarly strong, pH-dependent effect on bubble production in agarose gels containing aqueous soil extract (Table 3.8). In this case, 0.7 M 2-hydroxy-5-nitrobenzyl bromide at pH 5.0 reduced bubble production, as compared with the control buffered at the same pH, by over 95%, whereas at pH 8.0 the bubble reduction was 65%. Hence, the proteinaceous surfactants participating in the stabilization of long-lived microbubbles in aqueous soil extract must also contain tryptophan residues in their structure. At first this finding seemed difficult to reconcile with the work of Kojima (ref. 254,255) and, later, that of Bremner (ref. 263), who summarizes their findings on the tryptophan content of soils as follows: "Tryptophan, which is destroyed by acid hydrolysis but not alkali hydrolysis, could not be detected in alkali hydrolysates of soils" (ref. 263). However, Barman and Koshland (ref. 279) have pointed out that "basic hydrolysis occasionally

TABLE 3.8

Effect of 0.7 M 2-hydroxy-5-nitrobenzyl bromide (HNBB) on bubble production in buffered agarose gels (0.3% w/w, 0.27 ml) containing aqueous soil extract. (Taken from ref. 231.)

Gel treatment	No. of trials	Number of bubbles (mean $\pm$ S.E.M.)
pH 5.0 (0.1 M Na$_3$citrate) buffer	16	28.88 $\pm$ 1.93
pH 5.0 buffer + 0.7 M HNBB	16	1.25 $\pm$ 0.21
pH 8.0 (0.1 M Na$_3$citrate) buffer	16	38.06 $\pm$ 2.81
pH 8.0 buffer + 0.7 M HNBB	16	15.31 $\pm$ 2.23

66

gives accurate values [regarding tryptophan content] with model peptides and with some proteins, but poor results are usually obtained for inexplicable reasons... . The fact that alkaline or acid hydrolysis destroys tryptophan and produces unreliable assays indicates that a method [such as the use of 2-hydroxy-5-nitrobenzyl bromide] involving direct analysis of the protein is desirable" (ref. 271). Moreover, a pyridine derivative identified as α-picoline-γ-carboxylic acid, which Schreiner and Shorey (ref. 242) have suggested may be formed in soils by decomposition of tryptophan, has been isolated both from a Hawaiian soil by Shorey (ref. 241) and from other soils by Schreiner and Shorey (ref. 242).

3.6   CONCLUSIONS

In view of all the foregoing arguments, it is evident that the naturally occurring, largely hydrophobic surfactants which surround and stabilize long-lived gas microbubbles include proteinaceous compounds that contain, and whose surface activity depend upon, aromatic amino acid residues, particularly tryptophan. This belief is in complete accord with the well documented surface-active properties of a wide variety of proteins (ref. 52,53,194,283-307) and certain glycoproteins (ref. 308-320) at air/water interfaces. Lastly, the major finding in these particular experiments of protein-stabilized gas microbubbles in filtered aqueous extracts of forest soil may provide an explanation for the widespread occurrence of these long-lived microbubbles in nature: Specifically, humic substances, which are known to reversibly bind proteinaceous material (ref. 266,269) (that presumably may include surfactant-stabilized microbubbles) thereby forming complexes resistant to decomposition (ref. 252,253,265-267,269), "are among the most widely distributed natural products on the Earth's surface, occurring in soils, lakes, rivers, and in the sea" (ref. 321).

Chapter 4

CHARACTERISTIC GLYCOPEPTIDE FRACTION OF NATU-
RAL MICROBUBBLE SURFACTANT

Subsequent experimental work in this laboratory was aimed
at the systematic development of an efficient method for isolating
the proteinaceous surfactants, which help stabilize natural
microbubbles, from both commercial agarose powder and from
forest soil samples collected locally. Successful isolation of this
glycopeptide fraction was eventually achieved (ref. 322), and the
results obtained from an extended program of chemical analysis, to
further characterize and compare chemically these proteinaceous
surfactants from both natural substances, are described below.

4.1   ANALYTICAL METHODS

4.1.1   Isolation of microbubble glycopeptide surfactant from commercial agarose and forest soil

Microbubble glycopeptide surfactant was first successfully
isolated from research-grade, commercial agarose powder (Marine
Colloids SeaKem HGTP Agarose, lot no. 60808). The agarose powder was added (5% w/w) to distilled water (preheated at 100°C) and
dissolved by gentle shaking. After 30 min at 100°C, the agarose
solution was then added to a solution of acidified 95% ethanol/5%
water which was maintained at 55°C with constant stirring. The
original 95% ethanol solution had been adjusted to pH 2 by the
addition of HCl; after the addition of the predetermined volume of
5% agarose solution, the resulting solution represented an
ethanol/water volume ratio of 5:1 from which agarose readily
precipitated. The precipitated agarose was removed by filtration
with suction through a precleaned Gelman Acropor 3 μm (pore
diameter) membrane filter and both the filtrate and precipitated
agarose stored for later use. The ethanol-soluble glycopeptide surfactant remained dissolved in the filtrate even when stored (below

0°C) for several months. The precipitated agarose, from which the glycopeptide surfactant had been extracted, retained its capacity to form gels and these were later used in decompression tests (see below).

A similar method was subsequently developed to also isolate the microbubble glycopeptide surfactant from samples of aqueous soil extract, which were all prepared in the following manner. Approximately 500 g of (damp) surface soil, specifically kokokahi clay (A11and A12 horizons of this very fine, montmorillonitic clay) obtained from a forested area on windward Oahu, Hawaii, was added to 2 liters of distilled water and refrigerated. The mixture was gently swirled at intervals for one day and then all undissolved material allowed to settle during the second day. Thereafter, the supernatant was carefully removed and filtered with suction through a precleaned Gelman Acropor 3 μm membrane filter used in conjunction with a Gelman thick glass-fiber prefilter. Aliquots were then pipetted into separate glass vials and stored frozen for later use over the following 20 months in the soil extract experiments. When thawed for an experiment, none of the aliquots of soil extract contained any sediment. The thawed aqueous soil extract was light brown in color and upon addition of 4 volumes of 95% ethanol, a dark brown precipitate slowly formed. After 24 hr, the brown precipitate was removed by filtration with suction through a precleaned Gelman Acropor 3 μm membrane filter and the filtrate stored for later use. Again, the ethanol-soluble glycopeptide surfactant remained dissolved in the filtrate even when stored below 0°C.

4.1.2 Decompression tests with agarose gels

The pressure vessel, glass counting chambers, and their plexiglass holder used in the decompression experiments are all similar to those described earlier (ref. 153). As viewed through the plexiglass wall of the pressure vessel, the inside dimensions of each rectangular counting chamber are 15 mm wide and 6 mm along the line of sight.

Each chamber was filled with agarose solution to a depth of 7 mm. After gelation, the agarose samples were exposed to a fixed pressure schedule (i.e., 0 psig, then pressure increased 1 psig/sec ⇒ 85 psig (held for 20 hr at 21°C), then rapid decompression (within 15 sec) ⇒ 0 psig) using pure nitrogen gas. In control experiments, the 20-hr saturation period was found to be more than adequate to completely saturate the aqueous gel medium (ref. 139); however, only bubbles formed between the depths of 1 to 4 mm in the gel were counted. Therefore, the total volume of gel examined in each sample amounted to 0.27 ml. In this specified volume, all bubbles observed appeared randomly distributed and none of them were in contact with the walls of the glass counting chambers. (This is the expected distribution for bubbles that arise from free-floating long-lived gas microbubbles, which are also referred to as "stream nuclei" or "cavitation nuclei" in cavitation studies (ref. 323-325; see also Chapter 1).)

Research-grade agarose powder (Marine Colloids SeaKem HGTP Agarose, lot no. 60808), either with or without having undergone a protein extraction treatment performed in this laboratory (see above and Table 4.1), was used in the decompression tests. The agarose powder was added (1% w/w) to distilled water (preheated to 80°C) and dissolved by gentle shaking. After 20 min at 80°C, the agarose solution was transferred to a 60°C bath, from which it was pipetted into the acid-cleaned counting chambers.

4.1.3 Amino acid analyses of the isolated glycopeptide surfactant

The isolated microbubble glycopeptide surfactant, obtained from both commercial agarose powder and from forest soil extract, was prepared for (peptide-bond) hydrolysis by first evaporating the water/ethanol extraction solution to dryness on a rotary evaporator. The dried glycopeptide surfactant was redispersed in 5.7 N HCl containing 2% phenol and hydrolyzed in vacuum-sealed ampules for 24 hr at 110°C. Hydrolysates were centrifuged to remove salts and humin, and then evaporated to dryness with a stream of nitrogen. The hydrolyzed material was redissolved in water and evaporated to dryness again two times to ensure complete removal of HCl. The residue was taken up in 2.5-4.5 ml of pH 2 citrate buffer (ref. 326), centrifuged, and the supernatant filtered through a 0.3 μm millipore filter. Aliquots (0.45 or 1.0 ml each) of the filtrate were then charged on a programmed amino acid analyzer (Beckman Instruments, Model 120C). Amino acid values were calculated from normalized data to yield nearest integer values with each determination for the six most reliable amino acids, i.e., aspartic acid, glutamic acid, glycine, alanine, leucine, and lysine.

Nearest integer to the means was calculated from four determinations (Table 4.2). No corrections were made for destruction
(e.g., tryptophan) or incomplete hydrolysis.

4.1.4 Sodium dodecyl sulfate/polyacrylamide-gel electrophoresis

The isolated microbubble glycopeptide surfactant, obtained
both from commercial agarose powder and forest soil extract, was
subjected to sodium dodecyl sulfate/polyacrylamide-gel electrophoresis using several different procedures. The procedure used
for the gels shown in Fig. 4.1 was a modified version of the method
of Downer et al. (ref. 327). The water/ethanol extract containing
glycopeptide surfactant was concentrated on a rotary evaporator
and redispersed in a solution containing 3% sodium dodecyl sulfate
and 8 M urea. Some of the samples (see Fig. 4.1) were further
treated with 1% 2-mercaptoethanol. Electrophoresis was performed with 9.5% total acrylamide in a pH 6.4 Tris-acetate buffer
system (ref. 327) containing 1% (w/v) sodium dodecyl sulfate.
Samples were run for 15 hr at a constant current of 2.5 mA/gel.

The procedure used for the gels shown in Fig. 4.2 was that
of Swank and Munkres (ref. 328), except the samples were run for
18 hr at a constant current of 1.5 mA/gel and the protein-fixing
solution used after electrophoresis contained 4:5:1 methanol-water-
trichloroacetic acid.

The slab gel used to obtain the data given in Table 4.3 was
also prepared according to the method of Swank and Munkres (ref.
328), except the samples were run for 15 hr at a constant current of
15 mA/gel, followed by 3 hr at 50 mA/gel, and the protein-staining
solution was applied for a period of 12 hours after electrophoresis.

Finally, the slab gel used to obtain the data given in Table
4.4 was similarly prepared according to the method of Swank and
Munkres (ref. 328), except the samples were run for 16 hr at a
constant current of 15 mA/gel, followed by 3.5 hr at 50 mA/gel,
and the protein-staining solution was applied for a period of 24
hours after electrophoresis.

All gels used in Figs. 4.1 and 4.2 and Tables 4.3 and 4.4
were pre-electrophoresed in the appropriate running buffer for 12
hr before applying the sample. Samples were applied in amounts
ranging between 3-7 μg per cylindrical gel, and between 5-10 μg
per track on the slab gels. Staining of the samples following
electrophoresis was accomplished with either Coomassie Blue for

peptides (ref. 327) (all samples) or periodic acid-Schiff reagent for glycopeptides (ref. 329) (samples in Fig. 4.1 and 4.2 only). Molecular weight estimations for all glycopeptides were made according to the method of Weber and Osborn (ref. 330) using papain, cytochrome c, myoglobin CNBr fragments, and glucagon as standards.

4.1.5 Carbohydrate analyses of partially purified glycopeptide surfactant

The carbohydrate analyses commenced with an initial purification of the 80% ethanol-extracted glycopeptide surfactant (from forest soil). The 80% ethanol extract was forced (under a N_2 pressure of 65 psig) through a Nuclepore molecular filtration system (Model S43-70 stirred cell with a low-absorption membrane filter no. 1C7253), having a molecular weight cut-off of 1,000 daltons. The retentate, containing concentrated glycopeptide surfactant, was then diluted with (at least two more volumes of) high-purity absolute ethanol and the molecular filtration continued. The wash with absolute ethanol was repeated four more times. Preliminary tests revealed that the molecular filtration procedure effectively removed most of the low-molecular-weight contaminants, mainly inorganic salts and free sugars, from the original 80% ethanol extract containing microbubble glycopeptide surfactant. The purified solution was then evaporated to dryness on a rotary evaporator. The dried glycopeptide surfactant was redispersed in 2N HCl and (its glycosidic bonds) hydrolyzed in sealed ampules for 6 hr at 100°C (ref. 331). The hydrolysate was evaporated to dryness with a stream of nitrogen, redissolved in water, then filtered through a 0.45 μm millipore filter, and the filtrate evaporated to dryness two more times to ensure complete removal of HCl. The residue was then dissolved in 32 μl of water, and a 20-μl aliquot of the solution was charged on a programmed Tracor 985 high performance liquid chromatography (HPLC) system employing a Bio-Rad HPX-87 cation exchange column for monosaccharide separation (Fig. 4.3).

Thereafter, to remove suspected high-molecular-weight polysaccharide contaminants, a modified filtration procedure was developed. This particular procedure involved first passing the original 80%-ethanol extract of the microbubble glycopeptide surfactant through a 5,000-dalton cut-off molecular filter (Nuclepore membrane filter no. 1C7255). (Note that a 5,000-dalton cut-off was chosen because concurrent work in this laboratory with Sephadex column chromatography (see below) had shown the molecular weight of the microbubble surfactant to be less than 5,000 daltons. Accordingly, additional corroborative tests involving total amino acid (automated) analysis of the material retained by the 5,000-dalton cut-off filter showed this relatively high-molecular-weight material to be essentially devoid of amino acids.) The filtrate, which contained the glycopeptide surfactant, was then passed again through the same molecular filtration system, but this time outfitted with a membrane filter having a molecular weight cut-off of 1,000 daltons (filter no. 1C7253). The retentate, containing concentrated glycopeptide surfactant ($\approx$ 60 μg), was then diluted with high-purity absolute ethanol and the molecular filtration continued. The wash with absolute ethanol was repeated three more times. The purified solution was then evaporated to dryness and subjected to the same hydrolysis procedure as before. The hydrolyzed sample was then dissolved in 30 μl of water, and a 20 μl aliquot was charged on the same Bio-Rad HPX-87 cation exchange column for monosaccharide separation (Fig. 4.4).

4.1.6 Sephadex column chromatography of dansylated glycopeptide surfactant

In preparation for NH$_2$-terminal amino acid identification using the fluorescent label 5-dimethylamino-1-naphthalenesulfonyl (dansyl) chloride, the 80% ethanol-extracted glycopeptide surfactant (from forest soil) was further purified using both molecular filtration and Sephadex column chromatography. The Sephadex chromatography step was: 1) practicable using only 10-40 μg of labeled glycopeptide surfactant because of the high fluorescence of the small (270-dalton) dansyl label, and 2) had the advantage of providing an indication of the distribution of molecular weights of the individual glycopeptide molecules composing the microbubble surfactant. The purification first involved passing the 80% ethanol extract through a Nuclepore molecular filtration system (see above) having a molecular weight cut-off of approximately 1,000 daltons. The retentate, containing concentrated glycopeptide surfactant, was then diluted with high-purity absolute ethanol and the molecular filtration continued. The wash with absolute ethanol was repeated three more times. The purified surfactant solution was evaporated to dryness, and the glycopeptide surfactant (approximately 35 μg) was dansylated according to the method of Hartley (ref. 332). The freeze-dried, dansylated microbubble surfactant was dissolved in 100 μl of 50 mM-Na acetate buffer (pH 4.5) containing 4 M guanidine-HCl, and was chromatographed on a Sephadex G-25 "Fine" column (0.9 cm I.D. x 55 cm) equilibrated with the same buffer. The Sephadex G-25 column was eluted with the equilibrating buffer (at a flow rate of 1 ml/18 min), each 0.7 ml fraction combined with an added 0.8 ml aliquot of the same buffer, and the fractions analyzed for fluorescence immediately afterwards using an Aminco-Bowman Spectrophotofluorometer in conjunction with a Photomultiplier Microphotometer (American Instrument Co.). Activation of the dansyl label was performed at 254 nm, and the emitted fluorescent light was monitored at 450 nm.

Next, a column (1.5 cm I.D. x 35 cm) of Sephadex LH-20 was prepared, and then washed with 95% ethanol over a period of four days. Thereafter, the column was equilibrated with 95% high-purity ethanol/5% double glass-distilled water. Meanwhile, the dansylated microbubble surfactant contained in the fractions composing the central portion of the dansylated surfactant peak obtained previously from Sephadex G-25 chromatography (as shown by the cross-hatched portion in Fig. 4.5) were pooled and lyophilized. The resulting freeze-dried dansylated surfactant, now containing much salt contamination (mainly guanidine-HCl), was redissolved in 15 ml of high-purity absolute ethanol and then gradually evaporated. During evaporation, precipitated salt was removed by centrifugation until a final sample volume of 700 μl was obtained. The concentrated (and repurified) sample was applied to the Sephadex LH-20 column and was eluted with the equilibrating solvent mixture, at a flow rate of 1 ml/5 min. Each 0.5 ml fraction was combined with an added 0.8 ml aliquot of the same solvent mixture, and the fractions analyzed for fluorescence as described earlier. Fig. 4.5 shows that the dansylated glycopeptide surfactant molecules, obtained from the rather flat central portion of Peak I in Fig 4.5, were now separated on the Sephadex LH-20 gel matrix into five major peaks (Ia-Ie).

The dansyl analysis was thereafter directed toward obtaining an NH_2-terminal amino acid identification for each of these five major peaks (in Fig. 4.6). In each case, the NH_2-terminus was identified by two separate thin-layer chromatographic methods

(Table 4.5). The first was the three-dimensional procedure of Hartley (ref. 332) which employs the following three solvents: 1) 1.5% (v/v) formic acid; 2) benzene-acetic acid (9:1, v/v); 3) ethyl acetate-methanol-acetic acid (20:1:1, by vol.). The second method was the one-dimensional procedure of Morse and Horecker (ref. 333) and employed a modified solvent system containing benzene-pyridine-acetic acid (80:20:2) with 25 μl 2-mercaptoethanol added per 50 ml of solvent.

4.1.7 Edman degradation analyses

An entirely separate and corroborative NH_2-terminal analysis utilizing Edman chemistry (ref. 334) commenced with purification of the original 80%-ethanol extract of the glycopeptide surfactant (from forest soil). This ethanol extract was evaporated to dryness and then successively redissolved in anhydrous trifluoroacetic acid and distilled water separately (3 times for each solvent) to remove inorganic salts and some organic contaminants, respectively. The total amino acid composition of the glycopeptide surfactant was then redetermined at this second stage of purification, by the standard method for automated amino acid analysis described earlier, and no major changes in the amino acid ratios were noted (Table 4.6, 3rd column from left). Amino-terminal analysis of this same glycopeptide surfactant preparation, using the improved Edman degradation methods of Tarr (ref. 334) (see below), showed that at this second stage of purification the surfactant preparation still contained enough amino-containing (low-molecular-weight) contaminants to make any reliable NH_2-terminal (amino acid) identification of the glycopeptide surfactant improbable. Consequently, a third purification procedure was employed which involved passing the original 80%-ethanol extract of the microbubble glycopeptide surfactant through a Nuclepore molecular filtration system (see above) having a molecular weight cut-off of 1,000 daltons. The retentate, containing concentrated glycopeptide surfactant, was then diluted with high-purity absolute ethanol and the molecular filtration continued. The wash with absolute ethanol was repeated three more times. The total amino acid composition of the glycopeptide surfactant was then redetermined following this third purification procedure, and only minor changes in the amino acid ratios were noted (Table 4.6, 4th column from left). However, amino-terminal amino acid identification (ref.

334) of this particular preparation of the glycopeptide surfactant was now found to be far more promising. The Edman chemistry employed in the amino-terminal identification involved the use of a modified coupling reaction previously developed by Tarr (ref. 334). Following conversion to a phenylthiohydantoin (PTH), the PTH derivative of the NH_2-terminal amino acid(s) was detected via HPLC. With this third purification of the microbubble surfactant, the vast majority of the amino-containing contaminants, observed in the earlier NH_2-terminal identification attempts, had disappeared from preliminary HPLC chromatograms and only two dominant PTH-derivative peaks remained. In an effort to refine the analysis still further, a fourth purification procedure was employed which involved first passing the original 80%-ethanol extract of the microbubble glycopeptide surfactant through a Nuclepore molecular filtration system (see above) having a molecular weight cut-off of 5,000 daltons (membrane filter no. 1C7255). The filtrate, which contained the glycopeptide surfactant, was then passed again through the same molecular filtration system, but this time utilizing a membrane filter with a molecular weight cut-off of 1,000 daltons (filter no. 1C7253). The retentate, containing concentrated glycolpeptide surfactant, was then diluted with high-purity absolute ethanol and the molecular filtration continued. The wash with absolute ethanol was repeated three more times. The total amino acid composition of the glycopeptide surfactant was then redetermined following this fourth purification procedure, and again only minor changes in the amino acid ratios were noted (Table 4.6, 5th column from left). However, as expected, the amino-terminal identification (ref. 334) of this particular preparation of the glycopeptide surfactant was quite clear; only one dominant peak, having a retention time corresponding to the PTH-derivative (ref. 334) of alanine, appeared on the HPLC chromatogram (Fig. 4.7).

4.2   BIOCHEMICAL RESULTS

4.2.1   Protein extraction and bubble production in agarose gels
The proteinaceous (ref. 231) surfactants surrounding and stabilizing long-lived gas microbubbles, present in both carbohydrate (agarose) gels and forest soil extracts, have been successfully isolated from both of these natural sources. The isolation was first

76

TABLE 4.1

Bubble production following rapid decompression to atmospheric pressure of agarose gels (0.27 ml) saturated with N_2 at 85 psig. (Taken from ref. 322.)

Gel type	No. of trials	Number of bubbles (mean ± S.E.M.)
Agarose (lot no. 60808) before protein extraction	12	24.67 ± 2.95
Agarose (lot no. 60808) after protein extraction	36	0.08 ± 0.05

performed using research-grade agarose powder (from marine algae) in which the proteinaceous surfactant represents a trace contaminant that is incompletely removed during commercial manufacturing processes. Successful isolation of the microbubble surfactant from commercial agarose powder, using an acidified-ethanol extraction/millipore filtration procedure (see Section 4.1.1), was evidenced in part by the fact that the treated agarose powder was found to have lost over 99% of its bubble producing capacity in subsequent decompression tests (Table 4.1). A similar extraction method, using 80% ethanol and millipore filtration, was subsequently developed to also isolate the microbubble glyco-peptide surfactant from samples of aqueous soil extract (see Section 4.1.1).

4.2.2 Amino acid composition of microbubble glycopeptide surfactant

The isolated proteinaceous surfactants, obtained from commercial agarose powder and forest soil extract, were found to have extremely similar total amino acid compositions. Table 4.2 summarizes the amino acid values obtained from four determinations for each of the two surfactant preparations; it can be seen that the rather unusual amino acid ratios obtained (among 17 different amino acids) for the two separate cases closely resemble one another. Specifically, in both of these cases the relative amounts of the different amino acids identified (excluding tryptophan which is completely destroyed during acid hydrolysis) were as follows: glycine >> serine > aspartic acid (and/or aspara-

TABLE 4.2

Amino acid composition[a] of microbubble glycopeptide surfactant isolated from agarose powder and forest soil. (Taken from ref. 322.)

Amino acid	Surfactant from <u>agarose</u>	Surfactant from <u>soil</u>
Lysine	5 (6,5,3,4)	4 (3,4,3,7)
Histidine	1 (1,1,1,2)	1 (0,<1,<1,3)
Arginine	2 (2,1,2,1)	2 (1,1,1,3)
Aspartic acid[b]	9 (7,9,12,8)	10 (9,9,10,10)
Threonine	4 (4,3,4,5)	5 (6,6,5,4)
Serine	11 (13,8,8,16)	11 (15,9,9,9)
Glutamic acid[c]	9 (11,8,9,6)	9 (7,8,9,13)
Proline	1 (0,0,3,2)	4 (4,4,3,3)
Glycine	15 (15,16,13,14)	25 (29,26,25,22)
Alanine	7 (6,8,7,7)	10 (12,10,10,7)
Cysteine[d]	1 (0,0,2,1)	0 (0,0,1,1)
Valine	5 (4,7,4,4)	6 (6,6,6,5)
Methionine	0 (0,0,<1,<1)	0 (0,0,0,1)
Isoleucine	3 (2,3,4,2)	3 (3,3,3,4)
Leucine	5 (5,5,6,3)	5 (4,4,5,7)
Tyrosine	1 (2,1,1,1)	2 (3,2,2,2)
Phenylalanine	2 (3,1,2,2)	3 (4,2,3,2)

[a]Amino acid values are derived from normalized data to yield nearest integer values with each determination for the six most reliable amino acids, i.e., aspartic acid, glutamic acid, glycine, alanine, leucine, and lysine. Nearest integer to the means was calculated from four determinations, with the unaveraged data also given in parentheses. Hydrolysis in 5.7 N HCl for 24 h at 110°C. No corrections have been made for destruction (e.g., tryptophan) or incomplete hydrolysis.

[b]Value includes asparagine residues also.

[c]Value includes glutamine residues also.

[d] Cysteine as 1/2 cystine.

gine) ≈ glutamic acid (and/or glutamine) ≈ alanine > leucine ≈ valine ≈ lysine ≈ threonine > isoleucine ≈ phenylalanine ≈ proline ≈ arginine > tyrosine ≈ histidine ≈ cysteine > methionine (Table 4.2).

4.2.3 Molecular weight determinations by gel electrophoresis

The striking similarity between the proteinaceous surfactants obtained from agarose powder and forest soil extract was further demonstrated by sodium dodecyl sulfate/polyacrylamide-

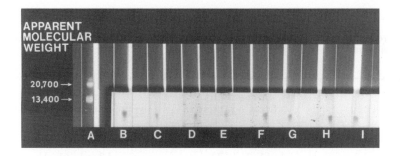

Fig. 4.1. Sodium dodecyl sulfate/polyacrylamide-gel electrophoresis of microbubble glycopeptide surfactant from commercial agarose and forest soil. Electrophoresis was performed with 9.5% total acrylamide (by a modified version of the method of Downer et al. (ref. 327)). The glycopeptide bands, stained with either Coomassie blue dye or periodic acid-Schiff reagent, were best visualized by illumination along the axis of the cylindrical polyacrylamide gels in conjunction with a black background. [Accordingly, to aid photography, the cylindrical gels were maintained upright in test tubes filled with appropriate storage solutions (with bubbles incompletely removed in 3 of the test tubes shown).] The movement of the glycopeptide material in the polyacrylamide gels during electrophoresis was toward the bottom of the photograph, and a tracking dye marker (which cannot be detected in the photograph) in each gel allowed proper alignment of the gels with respect to each other. Gel A in the test tube at far left contains two major (Coomassie blue-stained) bands representing two different protein standards: the upper band is papain (mol. wt. 20,700 daltons) and the bottom band is cytochrome C (mol. wt. 13,400 daltons). The remaining 8 gels in the photograph all contain microbubble glycopeptide surfactant. Gels B-E contain glycopeptide surfactant extracted (with acidified ethanol) from commercial agarose powder. Gels F-I contain glycopeptide surfactant extracted (with 80% ethanol) from Hawaiian forest soil. Furthermore, the glycopeptide material contained in gels D, E, H, and I were additionally treated with 1% 2-mercaptoethanol before application to the gels, whereas gels B, C, F and G did not receive this reduction treatment. Finally, gels B, D, F and H were stained with the protein-specific dye Coomassie blue, while gels C, E, G and I were stained with the carbohydrate-specific reagent periodic acid-Schiff stain. Migration characteristics of the glycopeptide surfactant in the test gels B-I were virtually identical, resulting in every case in a single, somewhat diffuse band corresponding to a calculated mol. wt. < 13,000 daltons (see text). (Taken from ref. 322.)

gel electrophoresis. Fig. 4.1 shows that the surfactants from both natural sources migrated as a single, somewhat diffuse band for the same distance during electrophoresis and, therefore, appear to have the same molecular weight. In addition, the migration distance was not affected by the addition of a sulfhydryl-reducing reagent (i.e., 1% 2-mercaptoethanol) in either case (Fig. 4.1). Hence, the surfactants from both natural sources appear to be linear peptides which do not contain (intermolecular or intramolecular) disulfide bonds; (note also the supporting (low) cysteine values listed in Table 4.2). Finally, both of the isolated surfactants appear to contain (covalently bound) carbohydrate since, in each case, the (8 M urea-treated) surfactant was successfully stained with either Coomassie blue dye (for peptides (ref. 327)) or periodic acid-Schiff reagent (for glycopeptides (ref. 329)) (Fig. 4.1). In view of all of the above data, it was clear that the microbubble surfactants isolated from agarose powder and forest soil extract were essentially indistinguishable. Consequently, all subsequent biochemical tests in this study utilized the microbubble glycopeptide surfactant from forest soil extract only, since the microbubble surfactant could be isolated in far greater quantity from this particular source.

The next biochemical test involved sodium dodecyl sulfate/polyacrylamide-gel electrophoresis of microbubble glycolpeptide surfactant from forest soil using much more dense gels, i.e., 12.5% total acrylamide (Fig. 4.2). These gels had been chosen because of the fact that the data obtained in Fig. 4.1 indicated that the microbubble surfactant had a molecular weight of less than 13,000 daltons. This finding was confirmed and extended in the electrophoretic study shown in Fig. 4.2, where the migration characteristics of the glycopeptide surfactant resulted in a calculated molecular weight of between 3,500-6,000 daltons. Again, the microbubble surfactant was successfully stained with either Coomassie blue dye (for peptides) or periodic acid-Schiff reagent (for glycopeptides) (Fig. 4.2).

Thereafter, approximate quantitative determinations of the carbohydrate content of the microbubble glycopeptide surfactant were made through the use of degradative enzymes. Table 4.3 summarizes the results from polyacrylamide (slab) gel electrophoresis of glycopeptide surfactant treated with β-N-acetyl-hexosaminidase, both alone and with endoglycosidase H. The

80

APPARENT
MOLECULAR
WEIGHT

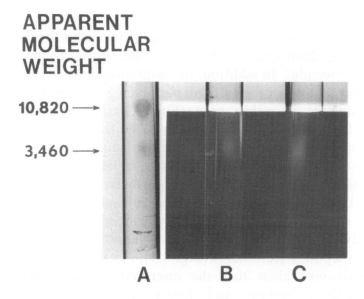

Fig. 4.2. Sodium dodecyl sulfate/polyacrylamide-gel electrophoresis of microbubble glycopeptide surfactant from forest soil. Electrophoresis was performed with 12.5% total acrylamide (by the method of Swank and Munkres (ref. 328)). Photographic conditions were identical to those described for Fig. 4.1. The movement of the glycopeptide material in the polyacrylamide gels during electrophoresis was toward the bottom of the photograph. Gel A in the test tube at far left contains two major (Coomassie blue-stained) bands representing two different protein standards: the upper band is myoglobin II/III CNBr fragment (mol. wt. 10,820 daltons) and the bottom band is glucagon (mol. wt. 3,460 daltons). The other two gels both contain microbubble glycopeptide surfactant from forest soil. Gel B was stained with the protein-specific dye Coomassie blue, while gel C was stained with the carbohydrate-specific reagent periodic acid-Schiff stain. Migration characteristics of the glycopeptide surfactant in the test gels B and C were virtually identical, resulting in each case in a single, somewhat diffuse band corresponding to a calculated mol. wt. between 3,500 - 6,000 daltons (see text). (Taken from ref. 322.)

apparent molecular weight of untreated glycopeptide surfactant was also determined again, with somewhat greater precision than obtained with the cylindrical gels, and was tentatively narrowed to between 4,400-6,000 (cf. below). In contrast, glycopeptide surfac-

TABLE 4.3

Enzymatic degradation[a] and slab gel electrophoresis of microbubble glycopeptide surfactant isolated from forest soil. (Taken from ref. 322.)

Sample[b]	Migration dist.(mm)	Apparent mol. wt.	Reduction of mol. wt. after enzyme action
Myoglobin	15	17,240	---
Myoglobin I/II	22	14,690	---
Myoglobin II/III	31	10,820	---
Myoglobin II	43	8,270	---
Myoglobin I	48	6,420	---
Myoglobin III	65	2,550	---
Microbubble glycopeptide surfactant (MGS)	49-55	4,400 - 6,000	---
MGS treated with β-N-acetylhexosaminidase	56-61	3,100 - 4,100	20 - 40%
MGS treated with β-N-acetylhexosaminidase & endoglycosidase H	56-62	3,000 - 4,100	20 - 40%

[a]The enzymatic conditions employed involved the use of 0.63 unit of β-N-acetylhexosaminidase (Sigma) both alone and with 0.01 unit of endoglycosidase H (Miles) in 220 µl of pH 4.75 Na acetate buffer (0.05 M) for 24 h at 20°C. The electrophoresis which followed was performed with a single slab (10 cm x 14 cm x 2 mm) of 12.5% total acrylamide (by the method of Swank and Munkres (ref. 328)).

[b]Myoglobin I/II, II/III, I, II, and III refer to the different cyanogen-bromide fragments of myoglobin which were used as molecular weight standards. Migration of all glycopeptide samples produced a more diffuse spot than observed with the standards.

tant treated with β-N-acetylhexosaminidase migrated a noticeably greater distance which corresponded to a calculated molecular weight of between 3,100-4,100 daltons (Table 4.3). Hence, the loss of (covalently bound) carbohydrate through the action of β-N-acetylhexosaminidase represented approximately 20-40% of the calculated molecular weight of the microbubble glycopeptide surfactant. This enzyme is reported by the manufacturer (Sigma) to liberate β-linked terminal N-acetyl-glucosamine and N-acetyl-galactosamine from a variety of natural and synthetic substrates. Interestingly, additional use of the degradative enzyme endoglycosidase H (Miles), which is known to act at interior sites within the (branched) carbohydrate chains of glycopeptides, resulted in no further reduction of molecular weight of the microbubble surfactant

82

TABLE 4.4

Exhaustive enzymatic degradation[a] and slab gel electrophoresis of microbubble glycopeptide surfactant (MGS) isolated from forest soil. (Taken from ref. 322.)

Sample[b]	Migration dist.(mm)	Apparent mol. wt.	Reduction of mol. wt. after enzyme action
Myoglobin	14	17,240	---
Myoglobin I/II	21	14,690	---
Myoglobin II/III	30	10,820	---
Myoglobin II	42	8,270	---
Myoglobin I	49	6,420	---
Myoglobin III	67	2,550	---
MGS treated with β-N-acetylhexosaminidase & endoglycosidase H	55-59	3,700 – 4,500	15 – 30%[c]

[a] The enzymatic conditions employed involved the use of 1.8 units of β-N-acetylhexosaminidase (Sigma) together with 0.025 unit of endoglycosidase H (Miles) in 400 μl of pH 4.75 Na acetate buffer (0.05 M) for 48 h at 20°C. The electrophoresis which followed was performed with a single slab (10 cm x 14 cm x 2 mm) of 12.5% total acrylamide (by the method of Swank and Munkres (ref. 328)).

[b] Myoglobin I/II, II/III, I, II, and III refer to the different cyanogen-bromide fragments of myoglobin which were used as molecular weight standards. Migration of the glycopeptide sample produced a more diffuse spot than observed with the standards.

[c] Calculation is based on a molecular weight of approximately 5,200 daltons for untreated microbubble glycopeptide surfactant, as determined in Table 4.3.

(Table 4.3). This same negative result for endoglycosidase H was found in a second test employing a higher concentration of both enzymes, and a much longer reaction period with the microbubble glycopeptide surfactant (Table 4.4).

4.2.4 HPLC determination of carbohydrate content

An effort was made to better quantitate this suggested preponderance of N-acetylglucosamine and/or N-acetylgalactosamine in the carbohydrate portion of microbubble glycopeptide surfactant by performing direct HPLC analysis of the monosaccharides contained in the surfactant. This analysis began with the (carbohydrate) hydrolysis of a partially purified preparation containing approximately 30 μg of the proteinaceous microbubble

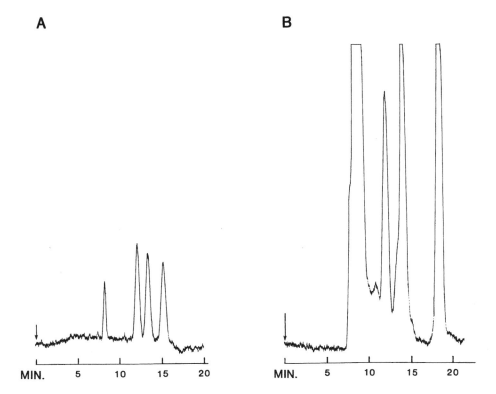

A   B

MIN.   5   10   15   20   MIN.   5   10   15   20

Fig. 4.3.   High performance liquid chromatography (HPLC) of the monosaccharides obtained from a partially purified preparation of microbubble glycopeptide surfactant from forest soil. Following hydrolysis (in 2 N HCl for 6 hr at 100°C) and filtration, the carbohydrate mixture was charged on a Bio-Rad HPX-87 cation exchange column. For comparison, part A shows the chromatogram (using the same HPLC column) of a standard solution, which contained 4 µg of each of three different monosaccharides (i.e., the last three peaks shown are glucose, xylose and fucose, in the order of increasing retention times). Part B shows the chromatogram obtained from hydrolysis of the partially purified (see text) microbubble surfactant (approximately 30 µg). All other experimental conditions were identical in the two cases, i.e., water eluent, 0.5 ml/min flow rate, 85°C, refractive index detector attenuation - 2x. (Taken from ref. 322.)

surfactant (see Section 4.1.5).   Fig. 4.3.(B) shows that actually several very large peaks were found, most of which were off-scale and contained obvious shoulders. Comparison with the standard (Fig. 4.3(A)) made clear that the hydrolyzed sample now contained far more free carbohydrate (by weight) than the total amount of proteinaceous microbubble surfactant (determined by amino acid analysis) initially added to the sample. This finding indicated that the microbubble surfactant preparation was contaminated with high-molecular-weight polysaccharides; concurrent work (see below, and Section 4.1) suggested that these polysaccharides could be removed selectively by prefiltering any subsequent surfactant preparation through a 5,000-dalton cut-off molecular filter. The added filtration step was instituted, and both the hydrolyzed > 5,000-dalton residue (Fig. 4.4(B)) and the hydrolyzed 1,000 < 5,000-dalton retentate (Fig. 4.4(C)) were analyzed separately. Comparison of Figs. 4.4(B) and 4.4(C) makes clear that the vast majority of the late-eluting monosaccharides, with retention times between approximately 14-19 min, were effectively removed by the additional 5,000-dalton cut-off filtration step before hydrolysis. Further comparison with a blank shown in Fig. 4.4(D) indicates that the very large, early composite peak (eluting at approximately 9 min) in both Fig4.4(B) and 4.4(C) is mostly artifact (see Section 4.1.5). Unfortunately, the large second peak, eluting at approxi-mately 12 min and also seen previously (Fig. 4.3(B)), persists in both Figs. 4.4(B) and 4.4(C), but does not appear in the chromatogram of the blank (Fig. 4.4(D)). This last finding indicates that the monosaccharides contained in this remaining second peak originate from contaminating polymers having molecular weights both above and below 5,000 daltons and, therefore, these carbohydrate polymers cannot be completely removed by simple molecular filtration methods. In view of the large quantity (i.e., > 70 µg) of monosaccharide contamination contained in this persisting second peak in Fig. 4.4(C) (cf. 4-µg standards shown in Fig. 4.4(A)), any further HPLC carbohydrate analyses of microbubble surfactant samples purified by the above-described molecular filtration methods would have been fruitless.

Nonetheless, direct verification of the presence of covalently bound carbohydrate in the proteinaceous microbubble surfactant was provided earlier by the enzymatic degradation experiments (see Section 4.2.3), and this finding has the added usefulness of providing a more accurate interpretation of the sodium dodecyl sulfate/polyacrylamide-gel electrophoretic data summarized earlier.

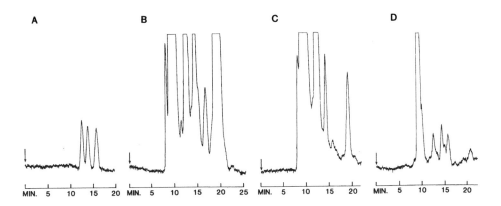

Fig. 4.4. High performance liquid chromatography of the monosaccharides obtained from a further purified preparation of microbubble glycopeptide surfactant. Following hydrolysis (in 2 N HCl for 6 hr at 100°C) and filtration, the carbohydrate mixture was charged on a Bio-Rad HPX-87 cation exchange column. For comparison, part A shows the chromatogram (using the same HPLC column) of a standard solution, which contained 4 μg of each of three different monosaccharides (i.e., glucose, xylose and fucose, in the order of increasing retention times). Part B shows the chromatogram obtained from hydrolysis of the >5,000-dalton residue collected from a preliminary filtration procedure, which was employed in the stepwise purification of the microbubble glycopeptide surfactant (see text for details). The filtrate, which contained the partially purified microbubble surfactant (60 μg), was then concentrated over a 1,000-dalton cut-off membrane filter (see text); this concentrated 1,000 < 5,000-dalton retentate was hydrolyzed in the same manner and the resulting chromatogram is shown in part C. Part D shows the chromatogram obtained from a blank, containing only the 2 N HCl used for the hydrolyses, which was subjected to the same preparative procedures. All other experimental conditions were identical in the four cases, i.e., water eluent, 0.5 ml/min flow rate, 85°C, refractive index detector attenuation - 2x. (See text for further discussion. Taken from ref. 322.)

This form of electrophoresis is not directly applicable to molecular weight determinations of glycoproteins. Glycoproteins and glycopeptides containing more than 10% carbohydrate behave anomalously during sodium dodecyl sulfate/polyacrylamide-gel electrophoresis when compared to standard proteins (ref. 335). The cause of this anomalous behavior is a decreased binding of sodium dodecyl sulfate per gram of glycoprotein as compared with standard proteins (ref. 335,336). The lower sodium dodecyl sulfate binding results in a decreased charge to mass ratio for glycoproteins versus standard proteins, a decreased mobility during electrophoresis, and thus a higher apparent molecular weight. However, with increasing polyacrylamide gel cross-linking, of the two factors (charge and molecular sieving) involved in electro-phoresis in sodium dodecyl sulfate/polyacrylamide gels, molecular sieving predominates and the anomalously high apparent molecular weights of glycoproteins decrease, approaching, in an asymptotic manner, values close to their real molecular weights (ref. 335). This is what one would expect since decreased mobility due to low sodium dodecyl sulfate binding is compensated by increasing gel sieving, i.e., increasing gel concentration (ref. 335) and/or em-ploying a high (8 M) urea concentration (ref. 328) (both of which are implemented to form Swank-Munkres gels). Since the molecular weight estimate for microbubble glycopeptide surfactant (determined first in Fig. 4.2 and then in more detail in Table 4.3) of between 4,400-6,000 daltons, or an average value of approximately 5,200 daltons, was obtained from experiments utilizing specifically Swank-Munkres gels, it is quite probable that 5,200 daltons is essentially equal to the (average) asymptotic minimal molecular weight for the untreated microbubble surfactant. However, additional methodology studies concerning sodium dodecyl sulfate/polyacrylamide-gel electrophoresis by other investigators has already shown that, "as a rough approximation, the asymptotic minimal molecular weight for a given glycoprotein appears to be about 1,000 daltons above its real molecular weight for every 10% of the glycoprotein represented by carbohydrate" (ref. 335). The data shown in Tables 4.3 and 4.4 suggest that somewhere between 15-40% of the microbubble glycopeptide surfactant is represented by (covalently bound) carbohydrate; actually, this percentage range is probably an overestimate and the 15% figure is closer to the true value since the electrophoretic mobility of the enzyme-treated (i.e., carbohydrate-poor) microbubble surfactant will be greater not only because of its decreased molecular size, but also because of its denser binding (ref. 335,336) of sodium dodecyl sulfate. Using the 15% figure as the probable carbohydrate content, the "corrected" (see above) molecular weight for the untreated microbubble glycopeptide surfactant is calculated to be approximately 4,000 daltons.

4.2.5 Gel-filtration column chromatography: Determination of average molecular weight and the NH_2-terminus

The lower, "corrected" <u>average</u> molecular weight for the microbubble glycopeptide surfactant is consistent with the findings from Sephadex column chromatography. Fig. 4.5 shows the distribution of surfactant molecules from dansylated microbubble surfactant on Sephadex G-25. Essentially the same elution profile was observed in a follow-up experiment using the same column. It was clear from each of these two experiments that the surfactant molecules were not excluded from the gel phase and, therefore, were very likely under 5,000 daltons (ref. 337). However, the broad peak (I) representing the dansylated microbubble surfactant was much closer than expected to the prominent salt peak (II, which included unbound dansyl label), i.e., the elution behavior of the dansylated surfactant on the column (calibrated beforehand with dansylated peptide standards) corresponded to a calculated (average) molecular weight of supposedly 2,000 daltons. Although the Sephadex G-25 column had been equilibrated with a buffer containing 4 M guanidine-HCl (ref. 338) and was eluted with the same buffer, hydrophobic interactions between the dansylated (largely hydrophobic) glycopeptide surfactant and the Sephadex gel matrix could not be ruled out entirely. Furthermore, many glycoproteins have been shown repeatedly to migrate anomalously on standard Sephadex G-type columns (ref. 339). Most of the anomalous effects on Sephadex G-25 were subsequently circumvented by switching to a Sephadex LH-20 column and a 95% ethanol/5% water eluent. For this experiment, the fractions composing the central portion of the single (dansylated) surfactant peak obtained previously from Sephadex G-25 chromatography (as shown by the cross-hatched portion of Fig. 4.5) were pooled, concentrated, and applied to the Sephadex LH-20 column (see Section 4.1.6). Fig. 4.6 shows that the dansylated glycopeptide surfactant molecules were now separated into five major peaks (I_a-Ie). The earliest of these peaks was eluted at the void volume (V_o) and, hence, contained surfactant molecules which were excluded

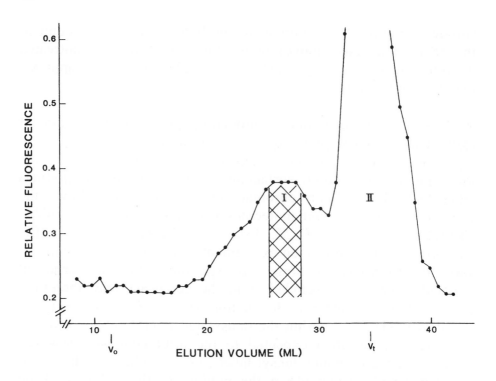

Fig. 4.5. Gel-filtration column chromatography of dansylated micro-bubble glycopeptide surfactant on Sephadex G-25. The column (0.9 cm I.D. x 55 cm) was equilibrated with 50 mM-Na acetate buffer, pH 4.5, containing 4 M guanidine-HCl. The column was eluted with the equilibrating buffer, at a flow rate of 3.3 ml per hour. The volume per fraction was 0.7 ml, and each fraction was combined with an added 0.8 ml aliquot of the equilibrating buffer before being analyzed for fluorescence. Activation of the dansyl label was performed at 254 nm, and the emitted fluorescent light was monitored at 450 nm. (Taken from ref. 322.)

from the gel phase. The other approximately 4/5 of surfactant contained in the remaining four peaks did succeed in entering the Sephadex LH-20 gel matrix, and the migration of some of these latter four peaks may actually have been somewhat retarded on the column (Fig. 4.6) due to either partition effects, reversible aromatic adsorption, and/or ion retardation effects using the 95% ethanol/5% water eluting solvent (ref. 340). In any case, it appears clear that

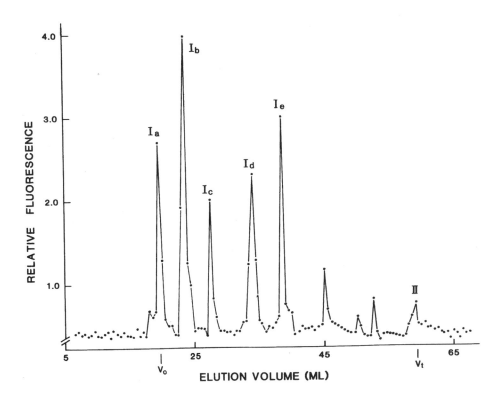

Fig. 4.6. Gel-filtration column chromatography of dansylated microbubble glycopeptide surfactant, obtained from the central portion of the surfactant peak from Sephadex G-25 chromatography (see Fig. 4.5), on a Sephadex LH-20 column (see details in the text). The column (1.5 cm I.D. x 35 cm) was washed and equilibrated with 95% ethanol/5% water for four days. The column was eluted with the equilibrating solvent mixture, at a flow rate of 1 ml/5 min. The volume per fraction was 0.5 ml, and each fraction was combined with an added 0.8 ml aliquot of the same solvent mixture before being analyzed for fluorescence. Activation of the dansyl label was performed at 254 nm, and the emitted fluorescent light was monitored at 450 nm. (Taken from ref. 322.)

the average molecular weight of the entire microbubble glycopeptide surfactant sample applied to the Sephadex LH-20 column (Fig. 4.6), and which represented the central portion of the single (dansylated) surfactant peak from Sephadex G-25 chromatography(Fig. 4.5), has a value slightly below the exclusion limit of the Sephadex LH-20 gel matrix. It is well documented that

TABLE 4.5

NH$_2$-terminal amino acid identification of the molecular sub-classes composing (dansylated) microbubble glycopeptide surfactant. (Taken from ref. 322.)

Peak from Sephadex LH-20 column chromatography (see Fig. 4.6)	Polyamide thin-layer chromatographic method[a] (3-dimensional)	Silica gel thin-layer chromatographic method[b] (one-dimensional)
Ia	alanine	alanine
Ib	alanine	alanine
Ic	alanine	alanine
Id	alanine	alanine
Ie	alanine	alanine

[a]Ref. 332.

[b]Ref. 333 with minor modification (see Section 4.1.6).

the exclusion limit of Sephadex LH-20 in pure ethanol is approximately 4,000 daltons (ref. 340). However, in the present experiments utilizing the 95% ethanol/5% water eluting solvent, the exclusion limit of the Sephadex LH-20 gel matrix would be expected to be slightly higher than 4,000 daltons, due to the greater swelling (ref. 340) of the gel in water as opposed to ethanol. Consequently, it is concluded that the _average_ molecular weight of the microbubble glycopeptide surfactant (as a whole) is very close to 4,000 daltons. This value is in excellent agreement with the "corrected" average value obtained from sodium dodecyl sulfate/polyacrylamide-gel electrophoresis, using the data from Tables 4.3 and 4.4 as described above (see Section 4.2.4).

Having determined the average molecular weight of the microbubble glycopeptide surfactant, subsequent experiments were concerned with evaluating the molecular heterogeneity across the five major peaks obtained from Sephadex LH-20 chromatography (Fig. 4.6). Were the five separate peaks (I$_a$-I$_e$) simply the result of microheterogeneity (ref. 341) within the carbohydrate portions of the constituent molecules, or did the different peaks comprise surfactant molecules having both different carbohydrate contents and different peptide chain lengths? Since the individual molecules within each surfactant peak already had been labeled (on their

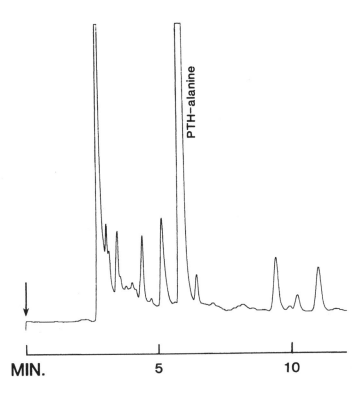

Fig. 4.7. High performance liquid chromatography of the NH$_2$-terminal amino acids obtained from an unfractionated (but further purified) preparation of microbubble glycopeptide surfactant (see details in the text). Following coupling, cleavage, and selective extraction of the phenylthiohydantoin derivatives of NH$_2$-terminal amino acids (using the improved Edman degradation methods of Tarr (ref. 334)), the PTH-amino acid mixture was charged on an Altex Ultrasphere ODS (5 μm) HPLC column calibrated for a variety of PTH-derivatives. The chromatogram obtained indicated the presence of several different PTH-amino acids, the largest peak by far being that for alanine (retention time ≈ 6 min). The computed area under this particular peak, which is off-scale, is more than 20 times the area under any of the other PTH-amino acid peaks. (Note that the initial, off-scale peak appearing in the chromatogram is an artifact resulting from side products formed by the Edman chemistry employed. Chromatogram taken from ref. 322.)

primary amino groups using the dansyl label), it could readily be determined whether (at least) the NH$_2$-terminal amino acids were the same for the surfactant molecules composing the five different peaks. The dansylated microbubble surfactant contained in the eluent fractions composing the different peaks (I_a-I_e) was pooled in each of the five cases, concentrated, and the NH_2-terminus identified by two separate thin-layer chromatographic methods (see Section 4.1.6). With each separate peak, it was found (Table 4.5) that the two methods yielded the same NH_2-terminus and only one. (Any secondary NH_2-terminal amino acids, if present, occurred at relative concentrations that were so low as to be undetectable.) Moreover, the same NH_2-terminal amino acid, specifically alanine (Table 4.5), was identified for all five major peaks (I_a-I_e) obtained from Sephadex LH-20 chromatography (Fig. 4.6). This finding clearly suggests that the separation of the microbubble glycopeptide surfactant, on Sephadex LH-20, into five distinct peaks is actually a reflection of microheterogeneity (ref. 341) within the carbohydrate portions, rather than different peptide chain lengths, of the constituent surfactant molecules. This suggestion is further supported by the fact that an entirely separate NH_2-terminal analysis, utilizing Edman chemistry (ref. 334), on the unfractionated (but further purified (see Section 4.1.7)) microbubble surfactant also resulted in the identification of only a single prominent NH_2-terminal amino acid, again specifically alanine, as detected by HPLC (Fig. 4.7).

4.3 REVIEW OF NATURAL-PRODUCT LITERATURE AND POSSIBLE ANIMAL SOURCES OF THE GLYCOPEPTIDE FRACTION OF MICROBUBBLE SURFACTANT

The prevalence of free glycoproteins in different water types throughout the environment, as well as the prominent surface activity of these molecules in such natural waters, have been exhaustively documented over a period of eight years by Baier et al. (ref. 342). These investigators suggest that such glycoproteins are of biological exudate origin. Other studies additionally indicate the involvement of glycoproteins specifically in bubble stabilization in diverse aqueous media, ranging from blood (ref. 320,343) to beer (ref. 51).

As concerns the physicochemical and biochemical data obtained in the present study, it becomes clear that the glycopeptide fraction of microbubble surfactant, isolated from agarose powder and forest soil extract, bears many strong similarities to other specific, surface-active glycoproteins which have been well-characterized chemically in the literature. For example, the appreciable solubility of the glycopeptide surfactant in 80% ethanol and its much reduced solubility in water is mimicked closely by the hydrophobic surface-active glycoprotein cerato-ulmin, a fungal toxin, which has also been found to surround and effectively stabilize microbubbles in aqueous media (ref. 344,345). However, the lower molecular weight, significantly different amino acid composition, and lack of disulfide cross-linking of the microbubble glycopeptide surfactant (see Section 4.2) make it easily distinguishable structurally from cerato-ulmin (ref. 346,347).

Interestingly, the total amino acid composition of the microbubble glycopeptide surfactant does, in fact, closely resemble the amino acid content of another surface-active glycoprotein, which is found in pulmonary lavage material and/or surfactant-secreting alveolar epithelial cells isolated from several different species of normal animals (ref. 309,310,314,318) (Table 4.6, 8th-10th columns from left). This 36,000-molecular-weight glycoprotein has also been found in human amniotic fluid (ref. 315) (i.e., from fetal lung surfactant), and in large amounts in lungs of humans with the chronic pulmonary disease alveolar proteinosis (ref. 308,311,312,319) (Table 4.6, 7th and 6th columns from left, respectively). In each case, the glycoprotein isolated displayed a total amino acid composition which was extremely similar to that determined for the microbubble glycopeptide surfactant (Table 4.6). Moreover, the pulmonary glycoprotein was also found to contain N-acetylglucosamine (ref. 313; cf. Section 4.2).

Additional work involving cyanogen-bromide cleavage of the 36,000-dalton glycoprotein isolated from pulmonary lavage material yielded a 5,000-dalton fragment (ref. 316,318). This (artificially produced) peptide fragment displayed an amino acid composition (ref. 210,212) (Table 4.6. 11th and 12th columns from left) which again closely resembled the rather unusual (ref. 309) amino acid molar ratios determined for the microbubble glycopeptide surfactant (Table 4.6). Furthermore, the similarity in molecular weight of this peptide fragment with that of the microbubble surfactant (see Section 4.2) suggests the possibility that the microbubble glycopeptide surfactant is essentially a partial degradation product of the surface-active, 36,000-dalton pulmonary

94

TABLE 4.6 Amino acid composition[a] of: purified surface-active glycopeptides from agarose powder, forest soil extracts and mammalian lungs; adsorbed protein films on tooth surfaces; humic substances from surface waters; and fulvic acids (FA) of organic soils. (Taken from ref. 322.)

Amino acid	Agarose powder[e]	Forest soil[e] (1st purif.)	Forest soil[f] (2nd purif.)	Forest soil[g] (3rd purif.)	Forest soil[f] (4th purif.)	Human lung[h]	Fetal lung[i]	Dog lung[j]	Chicken lung[k]	Rabbit lung[l]	Human lung[m] (CNBr fragment)	Rabbit lung[n] (CNBr fragment)	Tooth film[o] (from scaling)	Tooth film[p] (acid demin.)	Oregon river waters[q]	Georgia river waters[r]	Canada river waters[s]	Canada Lake and delta waters[s]	FA of muck bog[t]	FA of high-Cu muck bog[t]	FA of peat bog[t]	FA of high-Cu peat bog[t]	FA of a Podzol Bh horizon[t]
Lysine	5	4	5	5	4	4	4	2	4	4	4	4	7	4	3	4	2	2	2	2	2	2	2
Histidine	1	1	3	2	3	3	3	2	2	2	2	2	4	3	2	2	ND	ND	1	0	1	1	1
Arginine	2	2	3	3	3	6	6	3	6	6	4	4	4	4	1	0	ND	ND	1	ND	1	1	1
Aspartic acid[b]	9	10	9	10	8	10	10	12	8	8	8	8	7	9	14	14	14	14	15	17	14	15	14
Threonine	4	5	5	5	4	4	4	4	5	5	6	6	4	5	6	9	5	7	7	7	8	8	7
Serine	11	11	8	10	7	4	4	8	5	5	4	4	10	15	10	9	8	10	9	8	9	9	7
Glutamic acid[c]	9	9	12	14	11	10	10	13	9	9	8	8	13	13	10	8	10	9	9	9	9	9	14
Proline	1	4	1	5	3	7	7	1	6	6	4	4	2	6	4	5	4	4	4	3	6	6	4
Glycine	15	25	23	24	19	16	16	23	15	15	14	14	17	15	19	20	25	21	19	21	18	19	18
Alanine	7	10	12	6	7	8	8	7	7	7	8	8	7	8	12	11	10	11	12	14	13	13	12
Cysteine[d]	1	0	0	0	0	1	1	ND	2	2	2	2	1	0	ND	0	ND	ND	1	1	1	1	ND
Valine	5	6	4	5	4	4	4	2	6	7	6	6	4	4	5	5	7	6	6	6	7	7	7
Methionine	0	0	0	1	1	1	1	1	1	4	ND	ND	ND	1	4	1	2	3	0	ND	1	0	0
Isoleucine	3	3	2	4	3	5	5	1	4	4	4	4	3	3	2	3	1	1	3	3	3	3	3
Leucine	5	5	4	8	6	9	9	7	10	10	8	8	6	6	5	4	3	5	4	3	5	5	5
Tyrosine	1	2	1	3	2	3	3	8	3	3	2	2	2	3	0	2	2	1	3	3	ND	1	1
Phenylalanine	2	3	1	2	3	4	4	3	4	4	4	4	3	3	5	2	2	2	4	3	4	4	2

TABLE 4.6 (continued)

[a] Amino acid compositions are expressed as residues per 100 total amino acid residues, except for microbubble glycopeptide surfactant (i.e., in agarose powder and forest soil) where the values are derived from normalized data to yield nearest integer values with each determination for the six most reliable amino acids, i.e., aspartic acid, glutamic acid, glycine, alanine, leucine, and lysine. Nearest integer to the means was calculated from a minimum of three determinations (with exceptions noted below). Hydrolysis in 5.7-6 N HCl for 22-24 h, 110 $^\circ$C. No corrections have been made for destruction (e.g., tryptophan) or incomplete hydrolysis.

[b] Value includes asparagine residues also.

[c] Value includes glutamine residues also.

[d] Cysteine as cysteic acid, except for microbubble glycopeptide surfactant for which determined as 1/2 cystine.

[e] Ref. 322.

[f] Ref. 322. (Single determination.)

[g] Ref. 322. (Two determinations.)

[h] Ref. 311,312.

[i] Ref. 315.

[j] Ref. 314.

[k] Ref. 310.

[l] Ref. 318.

[m] Ref. 316.

[n] Ref. 318.

[o] Ref. 348.

[p] Ref. 350.

[q] Ref. 353.

[r] Ref. 351.

[s] Ref. 352.

[t] Ref. 354. (Single determination.)

glycoprotein, or even some other animal glycoprotein.

An alternate source of the microbubble glycopeptide surfactant could possibly be the adsorbed (salivary) glycoprotein films on animal teeth (ref. 348). Past studies by other investigators in which this adsorbed glycoprotein film (i.e., "acquired pellicle") was removed, either by careful scaling (ref. 348,349) or acid demineralization (ref. 350), and then subjected to acid hydrolysis have also yielded amino acid molar ratios (Table 4.6, 13th and 14th columns from left, respectively) which closely resemble those determined for the microbubble glycopeptide surfactant; furthermore, the acid hydrolysate of the tooth film was found to also contain both glucosamine and galactosamine (ref. 241), as already suggested for the proteinaceous microbubble surfactant from enzymatic degradation experiments and actually observed during the course of automated amino acid analysis (ref. 37). However, the solubility properties in (nonacidic) aqueous media of the glycoprotein molecules contained in the adsorbed tooth film are not known, nor has the molecular weight distribution of the constituent molecules been determined.

Both of the above-mentioned possible animal sources (in addition to any other biological exudate sources not yet detected) of the precursor glycoprotein(s) which give rise to the relatively low-molecular-weight microbubble glycopeptide surfactant (see Section 4.2) are in general agreement with the belief of Baier et al., noted above, that the glycoprotein material which they find in the organic films usually coating ambient air/water interfaces may be of biological exudate origin (ref. 342). Hence it is not surprising that the microbubble glycopeptide surfactant was found, in this study and in earlier studies (ref. 139,153,180,231) from this laboratory, to be present as a trace contaminant in commercial agarose powder since this carbohydrate substance is derived from marine algae (ref. 142-145). The presence of the microbubble surfactant in relatively large amounts in the surface layer of forest soil (see Sections 4.1 and 4.2) is also understandable in view of the widespread belief that "Humic materials are the product of a complex and ill understood sequence of reactions involving biological compounds excreted from living organisms or derived from decomposition of organisms after death" (ref. 226). The general sequence of events has been described by geochemists as follows: "Organic matter is preserved and stored mainly in the form of humic materials, which are resistant to microbial degradation. The water percolating through the swamp [or forest] environment carries these substances away in fairly high concentration. Of the rivers of the world, those originating in or flowing through tropical and subtropical vegetation will carry an increased load of this type of polymeric organic matter. …Condensation reactions of the above type would also explain the presence of amino acid residues found in the hydrolysates of river water organic matter" (ref. 351). It is useful to emphasize that the amino acid molar ratios determined for such hydrolysates of organic matter (which was resistant to further microbial degradation) from river, lake, and delta waters from many different areas of North America (Table 4.6, 15th-18th columns from left) are: 1) very similar across all water samples studied (i.e., a total of

31 sites (ref. 351-353)); 2) closely resemble the amino acid molar ratios reported for the fulvic acid fraction (which includes mostly the lower-molecular-weight humic molecules) of five different organic soils (ref. 354) (Table 4.6, 19th-23rd columns from left); and, most important, 3) also closely resemble the amino acid molar ratios determined for particularly the microbubble glycopeptide surfactant (Table 4.6, 1st-5th columns from left). Specifically, for practically all individual cases included within these three classes of samples (i.e., surface waters, fulvic acids of organic soils, and microbubble glycopeptide surfactant preparations), the relative amounts of the different amino acids identified (excluding tryptophan which is completely destroyed during acid hydrolysis) were as follows: glycine > serine, aspartic and glutamic acids, alanine > leucine, valine, threonine > lysine, isoleucine, phenylalanine, proline > arginine, tyrosine, histidine > cysteine, methionine (Table 4.6).

4.4   CONCLUDING REMARKS

In view of all the foregoing arguments, it appears likely that the microbubble glycopeptide surfactant is essentially a partial degradation product of larger, precursor glycoproteins, which may be of biological exudate origin and are widely distributed in the environment. Accordingly, the glycopeptide fraction of micro-bubble surfactant is actually composed of a small distribution of structurally similar (see Section 4.2) surface-active glycopeptides, rather than a single molecular species.

Chapter 5

ECOLOGICAL CHEMISTRY OF MICROBUBBLE SURFAC-TANT

Many natural waters and water extracts from soils often exhibit a yellowish to brown color owing to the presence of organic substances. The color in natural waters, according to Christman and Ghassemi (ref. 355), originates from the decay of forest vegetation. A detailed knowledge of the chemistry of these organic substances is needed to adequately account for important properties of the naturally occurring surfactants often found in forest soil (ref. 231,322). As emphasized earlier, the importance of these biological surfactants stems from the ability of some of them to stabilize gas microbubbles in both fresh and sea water over long periods of time (see Chapter 1). Accordingly, to understand the geochemical properties of the surfactant mixture surrounding natural microbubbles, a water-soluble extract from a Hawaiian forest soil, rich in microbubble surfactants (ref. 322), has been chemically characterized in the experiments described below (see also ref. 356).

A second objective of these experiments was to obtain additional biochemical data on the isolated microbubble surfactant mixture itself. This surfactant mixture was shown in Chapter 4 to contain low-molecular-weight glycopeptides of similar structure, which were invariably contaminated with a much greater quantity of oligosaccharide material (ref. 322). The low solubility in water (ref. 322) and oily nature of the microbubble surfactant mixture further suggested that lipids, previously unidentified, might also represent a major component of the mixture. In this chapter, we provide data on the biochemical heterogeneity of the microbubble surfactant mixture and identify the probable natural source of its characteristic glycopeptide fraction.

5.1   ANALYTICAL METHODS

5.1.1   Preparation of aqueous soil extract

The A_{11} + A_{12} horizons from a Udorthent soil, representative of the Kokokahi clay series, was used. This was of the montmorillonitic, isohyperthermic family, located at the island of Oahu, Hawaii (ref. 357).

Solutions were prepared using 2 kg of (damp) surface soil added to 10 liters of distilled water and refrigerated. The mixture was gently swirled at intervals for 1 day, and all undissolved materials were allowed to settle on the second day. The supernatant was filtered with suction through a precleaned Gelman Acropor 3-μm (pore diameter) membrane filter in conjunction with a Gelman thick glass-fiber prefilter. Two-thirds of the resulting filtrate was utilized for isolation of microbubble surfactant (see below), while the remainder was pipetted in 15-ml aliquots into separate glass vials and stored frozen until needed. The 15-ml aliquots were lyophilized to get the dry powder sample for analyses.

5.1.2   Elemental, infrared, and X-ray diffraction measurements

Elemental analyses were determined by the Canadian Microanalytical Service Ltd., Vancouver. The total Ca, Mg, P, Na, K, and Fe were determined by Dr. Carmen Mazuelos (Centro de Edafologia, CSIC, Sevilla, Spain) with a Perkin-Elmer Model 703 absorption spectrophotometer. The infrared spectrum of the lyophilized material was taken on KBr pellets (1 mg sample in 200 mg KBr) with a Perkin-Elmer Model 337 infrared spectrophotometer. X-ray diffraction analysis was carried out by Dr. Carmen Hermosin, Centro de Edafologia, in a Philips PW 1010 diffractometer using Ni-filtered, CuKα , radiation operated at 30 kV and 20 mA, scanning speed of 1°/min, time constant 2, and range of diffraction angles from 2 to 60°.

5.1.3   Pyrolysis mass spectrometry

The method has been described extensively in previous papers (ref. 358,359). The extract (1 mg) was suspended in methanol (1 ml) by mild ultrasonic treatment. From this suspension, 5-μl samples were applied to ferromagnetic coated wires and the solvent was evaporated under rotation. The coated wires were mounted in glass reaction tubes. The automated pyrolysis mass spectrometry system has been described in detail elsewhere (ref. 360). Briefly, it consists of a sample changing device, a Curie-point pyrolysis reactor with a high-frequency generator (Fisher Labortechnik GmbH, 1.5 kW, 1.1 MHz), a quadrupole mass spectrometer (Riber, QM17) with an ion counting detector, and a minicomputer (D-116, Digital Computer Controls). Pyrolysis was accomplished by inductive heating up to the Curie-temperature (510°C) of the ferromagnetic wire (Fe/Ni) within 0.1 sec. The total heating time was 0.8 sec. For normalization and graphic representation, peaks at 43 and 44 m/z were deleted in the case of the Hawaiian soil extract.

5.1.4 Isolation of microbubble surfactant

Isolation of the microbubble surfactant mixture itself from the aqueous soil extract followed the protein-extraction method used in earlier work (ref. 322) on forest soils and also commercial agarose from marine algae. Table 4.1 provides a typical example of the effectiveness of the protein-extraction method in removing surfactant-stabilized microbubbles from agarose gels (ref. 322) as evidenced by the drastic reduction in bubble production upon decompression of such gels. In the present study, the aqueous soil extract (see Section 5.1.1) was diluted with 4 volumes of 95% ethanol, after which a dark-brown precipitate slowly formed. After 24 hr, the precipitate was removed by filtration through a precleaned 3-μm membrane filter; the filtrate was then refiltered through a 5,000-dalton cutoff molecular filter (Nuclepore). This filtrate, which contained the microbubble surfactant, was concentrated over a 1,000-dalton cutoff molecular filter (Nuclepore). The retentate, containing concentrated surfactant, was then diluted with high-purity absolute ethanol and reconcentrated. The absolute ethanol wash was repeated three times, leaving a final retentate volume of 22 ml (containing approximately 550 μg of protein). Half of this volume was employed in a concurrent monolayer study (ref. 361), while the other half was concentrated by evaporation in preparation for column chromatography.

5.1.5 Gel-filtration column chromatography, amino acid analysis and carbohydrate determination

The concentrated microbubble surfactant sample (0.1 ml) was applied directly to a column (1.5 cm I.D. x 35 cm) of Sephadex

LH-20, which had been prewashed with 95% ethanol over a period of 1 week and thereafter equilibrated with 95% high-purity ethanol/5% water. The column was eluted with the equilibrating solvent mixture, at a flow rate of 1 ml/4 min. The volume per fraction was 0.65 ml, and each fraction was combined with an added 0.7-ml aliquot of the same solvent mixture before being analyzed on a Beckman Model 25 spectrophotometer at 230 and 280 nm.

Selected fractions were pooled and prepared for automated amino acid analysis by (complete) hydrolysis in 5.7 N HCl in evacuated, sealed ampules for 24 hr at 110°C. Analyses were performed on a Beckman 120C amino acid analyzer.

Other fractions were prepared for carbohydrate analysis by hydrolysis (of the surfactants' glycosidic bonds) in 2 N HCl in sealed ampules for 6 hr at 100°C (ref. 331). Monosaccharide analyses were performed on a programmed Tracor 985 HPLC system employing a Bio-Rad HPX-87C cation-exchange column for monosaccharide separation.

5.2   EXPERIMENTAL RESULTS

5.2.1   Abundant mineral content and characteristic IR absorption bands

Chemical characterization of the soil extracts yields, on an ash-free basis, 26.3% C, 4.6% H, 5.5% N, and 63.6% O + S (by difference). Ash content was 37.4% . The ash composition was 22.0% Ca, 8.0% Mg , 6.25% P, 2.06% Na, 1.51% K, and 1.75% Fe.

Although the application of IR spectroscopy to soil organic matter is somewhat limited by the nature of such material which contains numerous infrared active groups that lead to extensive overlapping of individual absorptions, several bands do appear as discrete or reasonably well defined in humic compounds (ref. 362,363). For example, in such IR studies, hydrogen bonded OH groups give a broad absorption band with a maximum around 3400 cm^{-1}, while carboxylate ions (COO^-) display bands at ca. 1620 and 1380 cm^{-1}, all of which are observed in the Hawaiian soil extract (Fig. 5.1). This spectrum is similar to those reported by Dormaar (ref. 364) for lyophilized defrost water obtained during spring thaw from Canadian Black Chernozemic A_h and A_p horizons, in which some of the bands were assigned to nitrates, sulfates, and carbon-

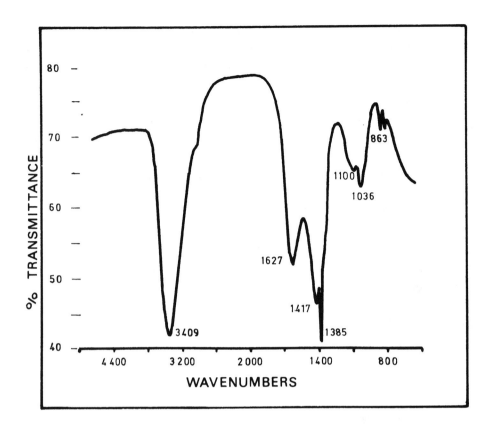

Fig. 5.1. IR spectrum of the Hawaiian soil extract. (Taken from ref. 356.)

ates. Spectra for pure sulfates (major band at ca. 1100 cm^{-1}), nitrates (major bands at ca. 1420 and 1385 cm^{-1}), and carbonates (major bands at 1420 and 875 cm^{-1}) showed absorption bands in agreement with those found in the Hawaiian soil extract. Both extracts showed that Ca and Mg were the main cations. Therefore, bands for organic components were probably masked by the inorganic contribution. The X-ray diffraction pattern, in Angstroms, was 2.82 (I/I$_0$ = 100), 1.99 (43), and 1.62 (12). The ASTM data for NaCl was 2.82 (100), 1.99 (55), and 1.62 (15). Therefore, NaCl was also present in the soil extract, although it was not evidenced in the infrared spectrum because of the lack of such absorption properties by this compound.

5.2.2 <u>Comparison of pyrolysis mass spectra for aqueous soil extract, fulvic acid, and water-soluble humic acid</u>

Fig. 5.2(A) presents the pyrolysis mass spectrum for the soil extract. In previous work (ref. 358,359,365) it was shown that complex organic materials like polysaccharides, proteins, lignins, and soil humic fractions have characteristic peaks yielding a typical pattern, which give preliminary information about the composition of the pyrolysis fragments. Thus, characteristic peaks for polysaccharides were observed at 60, 68, 82, 84, 96, 98, 110, 112, and 126 m/z, which were also present in the soil extract. They were shown to be related to acetic acid, furan, methylfuran, hydroxyfuran, furfural, furfuryl alcohol, methylfurfural, methoxy-methylfuran, and a typical pyrolysis fragment of polysaccharides with hexose and/or deoxyhexose units, respectively.

Alkenes were indicated by prominent peaks at 28, 42, 56, 70, and 84 m/z (ethene to hexene). They were probably derived from aliphatic acids, as evidenced by the fragment series at 43, 57, 71, 85 m/z, and a massive peak at 44 m/z (CO_2) indicating an extensive decarboxylation. (Note that 43 and 44 m/z are not shown in Fig. 5.2(A); see Section 5.1.3.)

The peaks seen at 67, 81, 95, and 117 m/z (Fig. 5.2(A)) are common for nitrogen-containing materials (e.g., proteins, peptides, amino acids) and are likely to represent (alkyl) pyrroles and indole. In addition, some peaks were related to aromatic compounds, for example, those at 92 (toluene), 94 (phenol), 108 (cresol), 120 (C_2-alkylbenzene and/or vinyl phenol), 122 (xylenol), 124 (guaiacol), and 138 m/z (methylguaiacol) (Fig. 5.2(A)). Guaiacol derivatives are typical pyrolysis products of lignins and lignin degradation products (ref. 366). Toluene, phenol, cresol, and xylenol arise from either aromatic amino acids, lignins, or humic substances.

Soil fulvic acid (Fig. 5.2(B)) shows dominant signals similar to those observed in a complex polysaccharide spectrum (ref. 359), with peaks at 68, 82, 84, 96, 98, 110, 112, 114, 126, and 128 m/z. In addition, nitrogen-containing fragments are low. Evidence of polysaccharide is also shown in the spectrum of a water-soluble fraction of humic acid (Fig. 5.2(C)), isolated by gel filtration (ref. 362). However, the presence of homologous ion series of sulfides (34, 48 m/z), pyrroles (67, 81, 95 m/z), pyridine (79 m/z), benzenes (78, 106 m/z), phenols (94, 108, 122 m/z), indole (117 m/z), and lignin-derived units (124, 138, 150 m/z) with variable intensity,

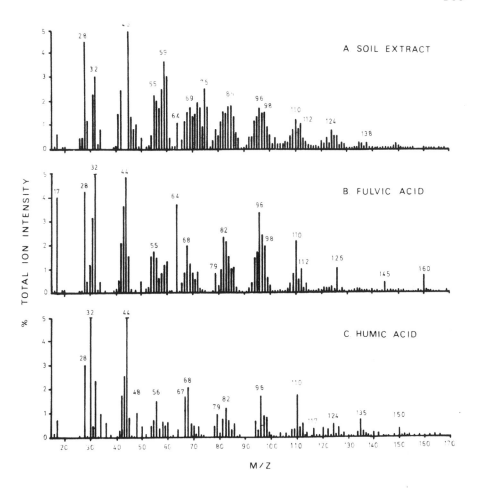

Fig. 5.2. Pyrolysis mass spectra: (A) Hawaiian soil extract; (B) fulvic acid from Typic Chromoxerert; (C) water-soluble humic acid from Typic Chromoxerert. (Taken from ref. 356.)

indicate the similarity with other soil humic fractions (ref. 359).

The above-mentioned strong indication of alkenes (derived from acyl lipids) by the pyrolysis mass spectrum of the soil extracts accords well with results from iodine-stained, thin-layer chromatography of the isolated microbubble surfactant mixture itself; these biochemical data indicated a high content of unsaturated acyl lipids (Reimer and D'Arrigo, unpublished data). This prominent lipid content, as well as the large proportion of oligosaccharide material, was again noted during concentration of the isolated microbubble surfactant mixture, prior to further purification via column chromatography in this study. During evaporation, a large quantity of whitish precipitate steadily formed. The whitish precipitate was subsequently found to contain carbohydrate and lipid material, the former determined by HPLC and the latter by thin-layer chromatography and nuclear magnetic resonance (NMR) spectroscopy (see also below).

5.2.3 Further purification of the microbubble surfactant mixture by gel-filtration column chromatography

The gel-filtration column chromatography of the con-centrated microbubble surfactant mixture, on Sephadex LH-20, resulted in the surfactant mixture being separated into three major peaks (Fig. 5.3). As in the previous biochemical study (see Chapter 4), most of the surfactant material (peaks I and II) eluted soon after the void volume (V_o); hence, the surfactant molecules succeeded in entering the LH-20 gel matrix, which has an exclusion limit in ethanol of approximately 4,000 daltons (ref. 340). However, there was a higher proportion of late-eluting (and possibly interacting (ref. 340)) surfactant material (peak III) from this preparative column chromatography than observed in the previous study. [This result and additional minor differences may well stem from the fact that although the forest soil used in both studies was taken from the same exact physical site, the actual dates of soil collection for the two cases differed by 1.5 years.]

5.2.4 Amino acid composition of the main glycopeptide subfrac-tion from microbubble surfactant

To obtain data on the heterogeneity of the glycopeptide fraction of microbubble surfactant, comparative amino acid analyses were performed on two of the major peaks obtained from gel filtration. From the ratio of absorbances at 230 and 280 nm (ref. 265) and the elution profile shown in Fig. 5.3, it appeared that peaks I and III would differ the most in amino acid composition and, therefore, these two peaks were selected for amino acid analysis. Peak I was sufficiently large to be divided into three equal aliquots and peak III into two equal aliquots for automated analysis. Peak II, which eluted closest (Fig. 5.3) to the dominant peak I and presumably was most similar in molecular composition to this large initial peak, was analyzed separately by HPLC for carbohydrate content.

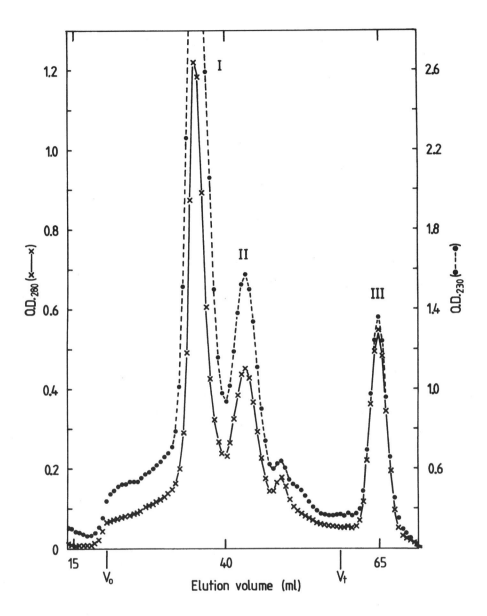

Fig. 5.3. Gel-filtration column chromatography of partially purified microbubble surfactant using Sephadex LH-20. The column (1.5 cm I.D. x 35 cm) was washed and equilibrated with 95% ethanol/5% water over a period of 1 week. The column was eluted with the equilibrating solvent mixture, at a flow rate of 1 ml/4 min. The volume per fraction was 0.65 ml, and each fraction was analyzed spectrophotometrically at 230 and 280 nm. (Taken from ref. 356.)

108

TABLE 5.1

Amino acid composition[a] of microbubble surfactant peaks obtained from Sephadex LH-20 chromatography (shown in Fig. 5.3). (Taken from ref. 356.)

Amino acid	Peak I determinations			Peak III determinations	
	1st	2nd	3rd	1st	2nd
Aspartic acid[b]	6.69	6.63	6.71	7.74	8.20
Threonine	7.64	7.70	7.69	6.76	6.09
Serine	10.72	11.03	10.91	11.72	10.27
Glutamic acid[c]	12.99	12.74	12.70	16.72	17.68
Proline	7.38	7.17	6.91	0	0
Glycine	15.53	15.80	15.25	22.17	21.59
Alanine	9.44	9.65	9.61	7.04	7.88
Cysteine[d]	0	0	0	0	0
Valine	9.71	10.03	10.01	5.69	4.94
Methionine	0	0	0	0	0
Isoleucine	5.31	4.74	4.90	3.14	3.30
Leucine	9.48	9.29	9.38	6.46	7.48
Tyrosine	1.19	1.02	1.35	3.53	3.39
Phenylalanine	2.04	1.92	2.12	2.61	2.62
Lysine	0.93	1.30	1.32	2.05	2.11
Histidine	0	0	0	0	0
Arginine	0.97	0.97	1.12	4.37	4.48

[a]Hydrolysis in 5.7 N HCl for 24 h at 110°C. No corrections have been made for destruction (e.g., tryptophan) or incomplete hydrolysis. Values for individual determinations are given in mole %.

[b]Value includes asparagine residues also.

[c]Value includes glutamine residues also.

[d]Cysteine as 1/2-cystine.

Table 5.1 lists the amino acid values obtained from the individual determinations for peaks I and III. Agreement between individual trials was excellent for the dominant peak I, and mostly very good for peak III where less proteinaceous material was available for analysis. Judging from the obvious absence of proline in peak III, it may well be the case that the material contained in this late-eluting (Fig. 5.3) peak, having a total amino acid composition (except for proline) somewhat similar to peak I, represents a partial degradation product (or several products) of the dominant, first peak. In any case, the very prominent, early peak (I) appears to have been quite homogeneous in its peptide composition, as judged from the excellent agreement of the separate amino acid analyses for this particular peak (Table 5.1). This finding may indicate the existence of a single, but yet widespread, type of glycoprotein precursor for the main glycopeptide subfraction from microbubble surfactant (see Table 5.2 and Section 5.3.4).

From the carbohydrate analyses of peak II, it was determined that 780 μg of polysaccharide (i.e., polymers of neutral monosaccharides with polymer mol. wt. of 1,000-5,000) were reversibly complexed with the estimated 19.5 μg of glycopeptide contained in this peak. Hence, the calculated carbohydrate:glycopeptide weight ratio is approximately 40:1.

5.3   BIOCHEMICAL/GEOCHEMICAL CONSIDERATIONS

5.3.1   Interaction of forest soil organic matter with abundant mineral content

Aqueous extracts from forest soils contain aliphatic organic acids and phenolic compounds, free or linked to condensed structures (ref. 367). It appears that the Hawaiian soil extract also includes these kinds of materials. Aliphatic acids are decarboxylated upon pyrolysis and the soil extract spectrum has peaks which predominate at 43 and 44 m/z (CO_2), which were deleted in order to normalize the spectrum for graphic representation. Furthermore, the carboxyl groups were most probably bonded to the inorganic constituents forming salts and chelates, explaining the high amount of ash found. Consequently, the ash and its catalytic effects, together with the relative sparsity of condensed structures, cause an extensive fragmentation of the Hawaiian soil extract by pyrolysis (Fig. 5.2(A)), in contrast to the soil humic fractions (see Figs. 5.2(B) and (C)) (ref. 359).

Polysaccharides probably constitute one of the most abundant fractions in soil organic matter and they are present as such or combined with the extracted humic fractions. As can be seen from the findings with soil extract and humic fractions (see Section 5.2), water-soluble materials are made up of a complex mixture of polysaccharides, proteins, phenolic and lignin-derived compounds, as well as other aliphatic materials. These compounds are present in various proportions, from which some of the oxygen-containing functional groups (carboxyls, phenolic and alcoholic hydroxyls, etc.) are capable of interacting with metal ions, metal oxides, metal hydroxides, and more complex minerals to form metal-organic associations of widely differing chemical and biological stabilities and characteristics. In this connection, Dawson et al. (ref. 368) pointed out that water-soluble compounds contain strongly chelated carboxylic acid groups playing a major role in the organic matter-metal interaction. Most of the water-soluble materials from forest soils are less than 5,000 daltons, and with them are associated most of the metals (ref. 368-370).

5.3.2 Dispersal of microbubble surfactants in natural waters

Similar to the above-described findings of Dawson et al., all of the microbubble surfactants which have been characterized biochemically to date from Hawaiian forest soil are known to be below 5,000 daltons (see Sections 4.2 and 5.2). The probable association of these rather hydrophobic (ref. 322) microbubble surfactants with the cations found in the Hawaiian soil extract, as suggested by our results and those of Dawson et al. (ref. 368), may well aid in the dispersal of the surfactants in natural waters. Accordingly, this oxygen group-metal interaction increases the ability of the microbubble surfactants to eventually reach the air-water interface of macroscopic bubbles in these waters. The surfactants contained in this water-soluble mixture would be expected to include glycopeptides, oligosaccharides, and acyl lipids, based upon the biochemical data obtained on isolated microbubble surfactant (see Section 5.2). Therefore, it is logical that the pyrolysis mass spectrum for the Hawaiian soil extract displays characteristic peaks for peptides, polysaccharides, and acyl lipids (see Section 5.2.2).

As the macroscopic bubbles in natural waters, loosely coated with surfactants, slowly shrink in size to become film-stabilized gas microbubbles (ref. 42,322), the oxygen group-metal interactions are probably replaced by hydrogen bonding between the various oxygen-containing functional groups of the surfactants in the insoluble monolayer. This transition appears likely since the biochemical findings (see below) indicate that the microbubble surfactant mixture actually represents a glycopeptide-lipopolysaccharide complex.

5.3.3 Bonding within the microbubble surfactant complex

The microbubble surfactant complex appears to be reversibly held together partly by hydrogen bonding, since either electrophoresis in 8 M urea-polyacrylamide gels or gel-filtration column chromatography (on Sephadex G-25) in eluting buffer containing 4 M guanidine-HCl was previously found (see Section 4.2) to be sufficient to monitor the behavior of the monomeric form of the glycopeptide fraction. Such hydrogen bonding between protein and carbohydrate in monomolecular films at an air/water interface has been described by others earlier (ref. 371,372). Nonpolar penetration also appears likely to be involved in formation of the microbubble surfactant complex since nonpolar side chains of the glycopeptides can penetrate into the lipophilic portion of the lipopolysaccharide material. Evidence for such reversible, nonpolar interactions is provided by the fact that, both in this study and the previous biochemical study (see Section 4.2), either molecular filtration (through a 5,000-dalton cutoff membrane filter) of the ethanol-solubilized microbubble surfactant or gel-filtration column chromatography (on Sephadex LH-20) in a 95% ethanol/5% water eluting solvent mixture also has been found sufficient to monitor the behavior of glycopeptide monomers.

The polysaccharide:glycopeptide weight ratio within the surfactant complex, as estimated from the chemical analyses of peak II (see Section 5.2.4), was found to be quite high, i.e., 40:1. Interestingly, this considerable ratio is still not sufficient to explain quantitatively the large areas encountered with the surface pressure-versus-area curves obtained for spread monolayers of the microbubble surfactant mixture at an air/water interface (ref. 361). Accordingly, additional (nondestructive) measurements involving NMR spectroscopy on aliquots from peak II (and peak I) beforehand revealed the presence of a very significant quantity of lipid remaining in this further purified, microbubble surfactant mixture as well (Rice and D'Arrigo, unpublished data).

5.3.4 Probable biological source of the glycopeptide fraction of microbubble surfactant

Despite the relatively small weight contribution of the glycopeptide fraction to the microbubble surfactant complex, this fraction has continually been found (ref. 322) to represent a reliable and characteristic component of microbubble surfactant

TABLE 5.2   Amino acid composition[a] of the main glycopeptide subfraction (peak I in Table 5.1) from microbubble surfactant and of the light-harvesting chlorophyll a/b-protein from higher plants and a green alga. (Taken from ref. 356.)

Amino acid	Peak I (mean)	Light-harvesting chlorophyll a/b-protein					
		Spinach beet[b]	Tobacco[b]	Oat[b]	Spinach[b]	Spinach[c]	Chlamydomonas[d]
Aspartic acid[e]	6.7	9.5	9.4	9.8	9.7	11.0	8.8
Threonine	7.7	3.1	3.0	3.8	2.6	3.1	5.0
Serine	10.9	4.4	4.2	3.3	3.4	5.5	3.6
Glutamic acid[f]	12.8	9.4	9.1	9.6	8.9	8.8	9.1
Proline	7.2	7.4	7.8	7.0	7.9	7.9	7.2
Glycine	15.5	13.1	14.0	13.1	13.8	13.6	12.5
Alanine	9.6	10.6	11.9	11.5	10.4	10.2	11.2
Cysteine[g]	0	0.5	---	---	---	---	0.6
Valine	9.9	6.8	6.6	7.1	7.1	6.3	4.4
Methionine	0	1.6	---	---	---	1.0	1.7
Isoleucine	5.0	4.6	4.7	4.7	4.2	2.6	4.4
Leucine	9.4	10.0	10.6	10.8	11.1	10.3	11.5
Tyrosine	1.2	2.6	2.3	2.8	2.2	2.8	3.3
Phenylalanine	2.0	5.8	6.7	6.3	7.2	6.4	6.8
Lysine	1.2	5.4	5.3	5.7	6.3	5.1	5.3
Histidine	0	1.1	0.9	1.0	1.4	1.4	1.5
Arginine	1.0	3.1	3.2	3.5	3.6	3.2	3.2
Tryptophan	---	1.1	---	---	---	1.0	---

[a]Hydrolysis in 5.7 N HCl for 24 h at 110°C. No corrections have been made for destruction (e.g., tryptophan) or incomplete hydrolysis. Values for individual determinations are given in mole %.

[b]Ref. 376.   [d]Ref. 378.   [f]value includes glutamine residues also.

[c]Ref. 377.   [e]value includes asparagine residues also.   [g]Cysteine as 1/2-Cystine.

preparations isolated from both marine algae and forest soils. The amino acid molar ratios shown in Table 5.1 (see Section 5.2.4), for peaks I and III, resemble those obtained earlier with less pure preparations (see Chapter 4) of the glycopeptide fraction, although the agreement among individual amino acid determinations is understandably far better in the present study. The agreement is particularly striking in the case of the main glycopeptide subfraction (peak I in Table 5.1), which may indicate the existence of a single type of protein precursor. An earlier search of the biochemical literature (see Section 4.3) suggested several animal sources for the protein precursor.

The literature search was now expanded and yielded the conclusion that another possible and more widespread precursor is the light-harvesting chlorophyll a/b-protein (LHCP), which is present in almost all photosynthetically grown higher plants and green algae (ref. 373-375). It is by far the major pigment-protein in these plants and accounts for 40-60% of the total chlorophyll (ref. 373,374).

Table 5.2 lists the amino acid molar ratios determined for LHCP from several plant sources, and compares these results with the mean values obtained for the main glycopeptide subfraction (peak I in Table 5.1) from microbubble surfactant. It can be seen from Table 5.2 that the amino acid composition of LHCP clearly resembles that of the main glycopeptide subfraction. Specifically, in both cases nonpolar residues represent a majority and near constant fraction (i.e., 59-62%) of the amino acid composition, with the relative amounts of such residues in practically all individual cases listed following the pattern: glycine > leucine, alanine, valine, proline > isoleucine, phenylalanine > methionine, tryptophan (Table 5.2). Accordingly, the glycopeptide fraction of microbubble surfactant may represent a degradation product of the light-harvesting chlorophyll a/b-protein, which is well known (ref. 373-375) to be extremely widely distributed in terrestrial, fresh-water, and salt-water environments (cf. ref. 379).

Chapter 6

SURFACE PROPERTIES OF MICROBUBBLE-SURFACTANT
MONOLAYERS

The biochemical work described in Chapter 5 indicated that
the microbubble surfactant mixture actually represents a
glycopeptide-lipid-oligosaccharide complex, which is reversibly
held together by both hydrogen bonding and nonpolar interactions.
The experiments described below were undertaken to examine in
detail the surface properties of the microbubble surfactant complex
at the air/water interface (ref. 361).

6.1 MODIFIED LANGMUIR TROUGH METHOD

The isolation of the microbubble surfactant mixture from
forest soil has been described in detail in preceding chapters.
Quantitative examination of the surface properties of monomolecular films of the isolated microbubble surfactant complex, at
an air/water interface, were carried out using a modified Langmuir
trough apparatus incorporating the surface tension method of
Padday et al. (ref. 380).

6.1.1 Surface pressure measurements with a cylindrical rod
A schematic diagram of the experimental set-up is shown in
Fig. 6.1. It employs a stainless steel rod of 0.5 cm diameter and
length 3 cm. The rod is hung by a thread from under a standard
bottom-loading balance (e.g., Shimadzu model EB-280) of
sensitivity 1 mg. This balance rests on a platform that can be
raised or lowered at speeds down to about 1.0 mm/min. On raising
the rod, the balance reading increases steadily, reaches a maximum
value (W_{max}) and then falls. The maximum is approached slowly
(i.e., over a period of approximately 10 sec for the last few mg) in
order to obtain equilibrium measurements. From the value of W_{max}
the weight of the rod in air (W_{rod}) is subtracted, and the surface

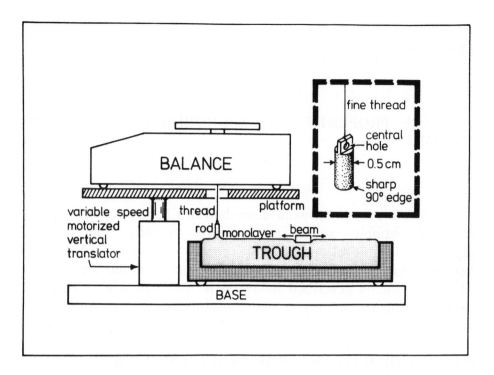

Fig. 6.1. Schematic drawing of Langmuir trough and balance for measuring surface pressures. (Taken from unpublished work by D'Arrigo, Israelachvili and Pashley.)

pressure of the film (Π) can then be read off directly from the graph in Fig. 6.2. The method is simple, sturdy, rapid and direct. It does not depend on the length of the rod, and requires <u>no</u> knowledge of the contact angle of the liquid attached to the rod; consequently, solids of different materials can be used, although stainless steel is recommended. The only experimental precautions required are: 1) that the diameter of the rod be accurately machined to 0.5 cm; 2) that its circular edge be sharp; 3) that it hangs vertically (i.e., straight) from the thread; and 4) that the rod is at least 2 cm away from the edge of the trough (D'Arrigo, Israelachvili and Pashley, unpublished results).

The theoretical basis of the method has been investigated in some detail by Padday et al. (ref. 380) who derived all the general equations needed for its application. Unfortunately, these general mathematical expressions tend to be complicated and rather incon-

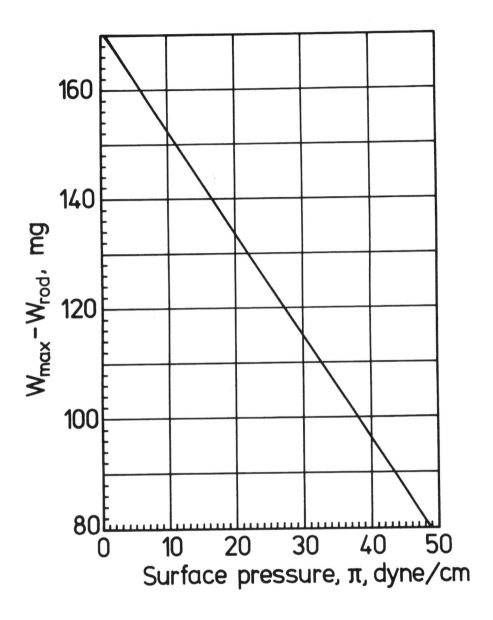

Fig. 6.2. Plot of surface pressure Π against maximum weight of 0.5 cm diameter rod at 21°C. W_{max} is the maximum weight measured, and W_{rod} the actual weight of the (dry) rod in air. For such a thin rod, the curve is almost a straight line, and to a high degree of accuracy is given by $\Pi = 0.54\,[\,171 - (\,W_{max} - W_{rod}\,)\,]$ dyne/cm. At 21°C, the surface tension of the water subphase is related to Π by $\gamma = (\,72.6 - \Pi\,)$ dyne/cm. (Taken from unpublished work by D'Arrigo, Israelachvili and Pashley.)

118

venient for widespread use by experimentalists in the biological sciences. However, using the experimental set-up described here, the detailed mathematical expressions were found to reduce in this case to a convenient, linear equation (Fig. 6.2).

6.1.2 Advantages of method when testing complex biochemical mixtures

The experimental method is based on the fact that if a cylindrical rod is pulled up through a liquid surface, it lifts up some liquid before finally detaching. As the rod is raised, its apparent weight, as measured by a balance, rises progressively to a maximum value and then falls prior to detaching. The maximum value gives the surface pressure <u>independent</u> of the contact angle of the liquid attached to the rod (ref. 380). (In contrast, the common use of a Wilhelmy plate, rather than a rod, carries with it the disadvantage (ref. 194) that it can be difficult to ensure a zero contact angle of the liquid on the plate with film-covered subphases; this problem can be a serious one (ref. 381).) With a balance that reads to ± 1 mg or better, it is possible with the set-up shown in Fig. 6.1 to determine pressures to at least ± 0.5 dyne/cm. Furthermore, the method is routinely found to provide remarkably stable and reproducible readings, even when testing monolayers of complex biochemical mixtures such as natural microbubble surfactant (see Chapter 5).

6.1.3 Langmuir trough apparatus and solutions

The trough itself measured 20 x 12 cm, was milled from a block of Teflon, and held approximately 750 ml of liquid. Monomolecular films were prepared on a subphase of (ultrapure) distilled water or on aqueous subsolutions containing varying concentrations of either NaF, HCl, NaOH, thiourea, or dimethyl sulfoxide (DMSO). Aliquots, between 50 and 250 µl, of the ethanol-solubilized microbubble-surfactant mixture were applied slowly to the surface of the subsolution from a Hamilton microsyringe. It was found unnecessary to allow the films to stand for more than 2 min after spreading before taking measurements. Furthermore, following compression or expansion, the surface pressure was observed to remain constant for periods of up to at least 10 min. All measurements were made at 20.0 ± 0.5°C.

6.2   SURFACE PRESSURE-AREA (Π-A) CURVES

6.2.1   Initial compression-expansion cycle

As can be seen from Fig. 6.3, it was found that the partially purified, microbubble-surfactant mixture does in fact form stable monomolecular films at an air/(distilled) water interface. During the first compression-expansion cycle a minor degree of hysteresis was observed, but this effect was essentially absent during recompression (Fig. 6.3) and is probably due to the presence of various contaminants in the microbubble-surfactant mixture (see Section 6.3). It was further found that these microbubble-surfactant monolayers remain quite insoluble (cf. Section 6.1.3) when highly compressed, i.e., up to measured surface pressures of 24 dyne/cm.

6.2.2   Effect of salt concentration, pH, and selected nonelectrolytes

Fig. 6.4 demonstrates that equally stable monolayers could be formed on subphases containing salt concentrations up to at least 1 M (NaF) or having pH's anywhere between 1.1 (0.1 M HCl) and 12.3 (0.1 M NaOH). To avoid any artifacts arising from contaminants in the microbubble-surfactant mixture which are apparently selectively desorbed from the monolayer during the initial compression phase (cf. Fig. 6.3 and Section 6.3), the data plotted in Fig. 6.4 include only Π-area measurements made during the expansion phase following an initial compression to at least 23 dyne/cm. The presence of thiourea (9% w/w) in the subphase resulted in a greatly expanded Π-A curve being obtained for the microbubble-surfactant mixture. An equimolar concentration of the nonelectrolyte DMSO was then tested as a control and, accordingly, the resulting Π-A curve was not significantly different from the group of curves obtained on subphases of either distilled water, 0.1 M HCl, 0.1 M NaOH, or 0.1 M NaF. However, use of a much higher concentration of NaF (i.e., a nearly saturated solution of 1.0 M) resulted in a Π-A curve for the microbubble-surfactant mixture which was essentially identical to the curve obtained on the subphase containing 9% thiourea (Fig. 6.4, and see Section 6.4).

120

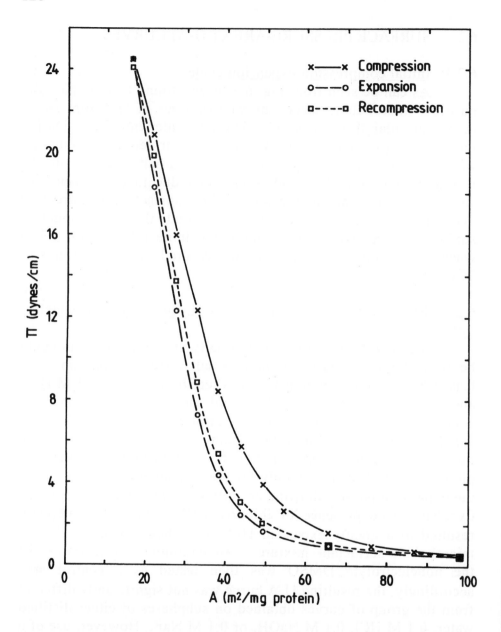

Fig. 6.3. Surface pressure-area (Π-A) curves for a microbubble-surfactant monolayer spread at the air/(distilled) water interface. (Note that area is expressed as m^2/mg protein in this figure and Figs. 6.4 and 6.5, which is also equivalent to m^2/110 mg of microbubble-surfactant mixture. See text for further discussion. Taken from ref. 361.)

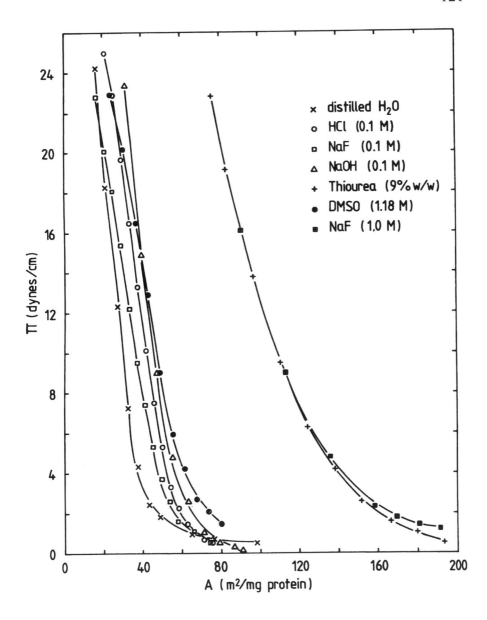

Fig. 6.4. Surface pressure-area (Π-A) curves for microbubble-surfactant monolayers spread on various aqueous subphases. (Taken from ref. 361.)

6.2.3 Π̲A̲-̲Π̲ ̲p̲l̲o̲t̲s̲
Fig. 6.5 displays the ΠA-Π plots for microbubble-surfactant

monolayers, in the region of low surface pressures (i.e., ≤ 1.8 dyne/cm), using the same seven subphases employed for the experiments summarized in Fig. 6.4. The monolayer data obtained on the distilled-water subphase yielded the largest ΠA intercept, while the 0.1 M NaF, 0.1 M HCl, 0.1 M NaOH, and 1.18 M DMSO data all yielded an identical and much lower ΠA intercept that was 1/6 of the value obtained using the distilled water subphase. Subphases containing either 9% (w/w) thiourea or 1.0 M NaF produced monolayer data that resulted in a single, intermediate ΠA intercept, which was ¼ of the value obtained with the distilled water subphase.

6.3 SELECTIVE DESORPTION FROM COMPRESSED MONOLAYERS

The biologically derived, ethanol-solubilized microbubble-surfactant mixture was observed to spread rapidly at an air/water interface, yielding a monomolecular film (Fig. 6.3). The spreading of this partially purified, surfactant mixture may well have been aided (cf. Chapter 5) by its very appreciable oligosaccharide and acyl lipid content, detected in other studies using high-performance liquid chromatography (ref. 322) and pyrolysis mass spectrometry (ref. 356), respectively. Additional results from NMR measurements on the collected monolayer material (see Chapter 7) indicate that almost all of the oligosaccharide material is, thereafter, selectively desorbed from the microbubble-surfactant monolayer during the initial compression phase (cf. Fig. 6.3). (Unlike the reproducible hysteresis noted for natural-lung-surfactant monolayers (ref. 382-384), the initial hysteresis observed with microbubble-surfactant monolayers during the first compression-expansion cycle does not recur in subsequent cycles; consequently, the material squeezed out of microbubble-surfactant monolayers during the initial compression phase evidently does not re-enter the monolayer upon subsequent expansion.) Hence, while the partially purified, microbubble-surfactant mixture has already been determined to represent a glycopeptide-lipid-oligosaccharide complex (see Chapter 5), it appears that, following initial compression of the microbubble-surfactant monolayer, essentially only glycopeptide-acyl lipid complexes remain at the air/water interface. Similarly, one would expect that when surfactant-coated

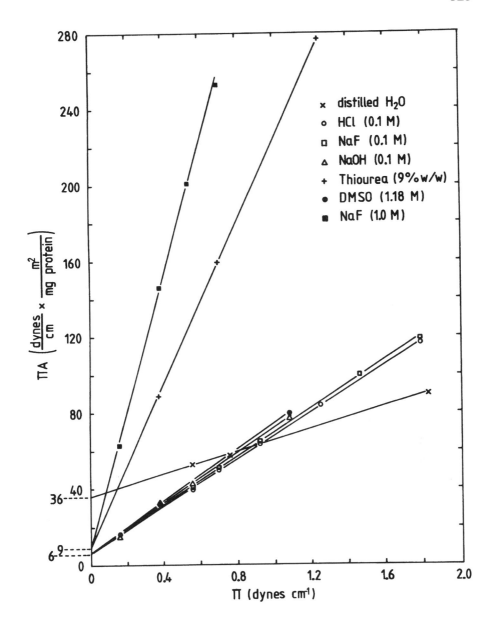

Fig. 6.5. Plots (ΠΑ-Π) for microbubble-surfactant monolayers spread on various aqueous subphases. (Taken from ref. 361.)

macroscopic bubbles in natural waters shrink in size (with decreasing speed due to a decreasing rate of dissolution of the contained gas (ref. 115,118-120)) to form film-stabilized microbubbles (ref. 8,15-17,42,50,322), the result is a population of long-lived gas microbubbles which in each individual case is surrounded by a densely packed, insoluble monolayer of glycopeptide-acyl lipid complexes.

Interestingly, this view concerning the biochemical composition of the stable microbubble-surfactant monolayer, which results following initial compression, is further supported by results from early monolayer experiments using microbubble-surfactant preparations of graded purity. It was found in the course of these initial monolayer experiments (D'Arrigo, unpublished data) that less pure microbubble surfactant preparations containing approximately the same amount of glycopeptide (determined by amino acid analysis and having a characteristic, average mol. wt. ≈ 4,000 (ref. 322)), from which the ethanol-soluble material above 5,000 daltons (i.e., polysaccharides with strongly adsorbed acyl lipids) had deliberately not been removed, displayed essentially the same family of Π-A curves as the further purified surfactant material used in the experiments of Fig. 6.3. It was therefore clear that the ethanol-soluble material larger than 5,000 daltons possessed little (if any) surface activity and, accordingly, that most of the surface activity was inextricably associated with the glycopeptide fraction of the microbubble surfactant mixture. At the same time, the area measurements per milligram protein (cf. Fig. 6.3) made it apparent that the glycopeptide:acyl lipid weight ratio in the stable, microbubble-surfactant monolayer was quite low (cf. Section 6.5).

6.4 BONDING WITHIN COMPRESSED MICROBUBBLE-SURFACTANT MONOLAYERS

Fig. 6.4 shows the Π-A curves obtained for microbubble-surfactant monolayers on a variety of aqueous subphases. In order to compare Π-A measurements for monolayers which would contain essentially only glycopeptide-acyl lipid complexes, the data plotted in Fig. 6.4 include only Π-A measurements made during the expansion phase (following an initial compression to at least 23 dyne/cm). Judging from the fact that subphases of either distilled water, 0.1 M HCl (pH 1.1), 0.1 M NaOH (pH 12.3), or 0.1 M NaF all yielded rather similar Π-A curves (Fig. 6.4), it appears clear that electrostatic interactions play only a very minor role in the formation of the glycopeptide-acyl lipid complexes contained in the microbubble-surfactant monolayer. This finding agrees well with results from separate biochemical work which indicated that the glycopeptide-lipid-oligosaccharide complexes, which constitute partially purified preparations of the microbubble surfactant mixture, are reversibly held together by hydrogen bonding and nonpolar interactions (see Chapter 5).

The suspected major role of hydrogen bonding was clearly supported by the striking finding that the presence of thiourea (9% w/w) in the subphase resulted in a greatly expanded Π-A curve being obtained with the microbubble-surfactant monolayer (Fig. 6.4). At room temperature, 9% thiourea is known to very effectively disrupt hydrogen bonding within complex molecules (ref. 385). This compound was chosen over other hydrogen-bond-breaking reagents, such as 8 M urea and 4 M guanidinium-HCl which were utilized in an aforementioned biochemical study (see Chapter 5), because of the fact that only thiourea displays almost none of the often parallel tendency of such reagents to solubilize peptides (ref. 385). From the large areas encountered when making the Π-A measurements with a subphase of 9% thiourea, it appeared evident that the reagent successfully broke (accessible) hydrogen bonds within the glycopeptide-acyl lipid complexes (thereby allowing their areas to increase) which reside in the microbubble-surfactant monolayer. This belief was confirmed in control experiments utilizing an equimolar concentration (i.e., 1.18 M) of DMSO, a water-soluble nonelectrolyte with very little or no capacity for breaking hydrogen bonds (cf. ref. 129). Accordingly, it can be seen from Fig. 6.4 that the Π-A curve obtained with a subphase of 1.18 M DMSO was not significantly different from the group of curves obtained on subphases of either distilled water, 0.1 M HCl, 0.1 M NaOH, or 0.1 M NaF. However, a concentration of NaF much closer (e.g., 1.0 M) to that used for thiourea would be expected to also break hydrogen bonds within the glycopeptide-acyl lipid complexes contained in the microbubble-surfactant monolayer, simply because the fluoride ion has been shown to be capable of forming rather strong hydrogen bonds with hydroxyl groups (ref. 386,387). As expected, Π-A data obtained with a subphase of 1.0 M NaF was essentially indistinguishable from the 9%-thiourea data (Fig. 6.4).

6.5 GLYCOPEPTIDE:ACYL LIPID AREA RATIO AND ASSOCIATION OF COMPLEXES WITHIN MONOLAYERS

An effort was also made to determine the approximate glycopeptide:acyl lipid area ratio in the stable microbubble-surfactant monolayer. For the purpose of this calculation, detailed Π-A measurements were obtained with (previously compressed) microbubble-surfactant monolayers in the region of low surface pressures (i.e., Π ≤ 1.8 dyne/cm). The data was organized in the form of ΠA-Π plots, where area was expressed in square meters per milligram protein (Fig. 6.5). A simple thermodynamic derivation (ref. 129) demonstrates that 1 mole of the spread substance will yield, at 20°C, an intercept of 2,438 at zero pressure when ΠA is plotted against Π and the area of the film is expressed in square meters per milligram; the molecular weight of the spread material is, at 20°C, equal to 2,438 divided by the intercept value of ΠA (ref. 129,149). The correct ΠA value to be used in the case of a (previously compressed) microbubble-surfactant monolayer is actually only a fraction of the intercept value shown in Fig. 6.5 since area is expressed, for ease of biochemical measurement, in square meters per milligram of protein (i.e., glycopeptide) rather than square meters per milligram of spread substance (i.e., glycopeptide + acyl lipid). Accordingly, 2,438 divided by the "corrected" ΠA value equals the glycopeptide molecular weight; since the latter is already known from preceding biochemical studies (see Chapters 4 and 5) to be approximately 4,000 daltons, the "corrected" ΠA value is 2,438/4,000 ≈ 0.61 . With regard to the data obtained on a distilled water subphase, it can be seen from Fig. 6.5 that the uncorrected intercept value is 36 which, in turn, yields a glycopeptide:acyl lipid area ratio of approximately (0.61:36 or) 1:60.

Interestingly, the ΠA-Π plots for microbubble-surfactant monolayers on subphases other than distilled water provide indirect evidence for induced association of the glycopeptide-acyl lipid complexes themselves. It can be seen from Fig. 6.5 that the monolayer data obtained on subphases of 0.1 M NaF, 0.1 M HCl,

0.1 M NaOH, or 1.18 M DMSO all yielded an identical ΠA intercept that was 1/6 of the value obtained using the distilled water subphase. The simplest interpretation of this observation is that under these experimental conditions the individual glycopeptide-acyl lipid complexes associate into hexamers, i.e., two-dimensional macrocomplexes comprising six glycopeptide-acyl lipid complexes. In contrast, subphases containing either 9% (w/w) thiourea or 1.0 M NaF produced monolayer data that resulted in a single ΠA intercept that was now ¼ of the value obtained with the distilled water subphase (Fig. 6.5). Since both subphases would be expected to bring about breaking of the (accessible) hydrogen bonds within the glycopeptide-acyl lipid complexes and thereby significantly increase the area of the individual complexes (cf. Fig. 6.4), the geometric packing constraints operating in the plane of the microbubble-surfactant monolayer may now favor the formation of tetramers rather than hexamers.

6.6   CONCLUSIONS

The long-lived gas microbubbles present in natural waters are surrounded and stabilized by a densely packed, monomolecular film of biologically derived surfactants. The microbubbles are probably formed from shrinkage of surfactant-coated macroscopic bubbles, and various data indicate that most of the carbohydrate material is selectively desorbed from the microbubble-surfactant monolayer during this initial compression phase (cf. ref. 388,389). The result is a stable, insoluble monolayer containing tightly packed, glycopeptide-acyl lipid complexes (cf. ref. 388,389; see also 390-393), which have been shown to be held intact primarily by hydrogen bonding. The glycopeptide:acyl lipid area ratio within the stable microbubble-surfactant monolayer is approximately 1:60, and association of the glycopeptide-acyl lipid complexes themselves within the monolayer can be induced by changes in the solute composition of the aqueous phase.

Chapter 7

STRUCTURE OF PREDOMINANT SURFACTANT COMPO-
NENTS STABILIZING NATURAL MICROBUBBLES

In the series of microbubble experiments (ref. 394) included
in this chapter, the actual film material, contained in compressed
microbubble-surfactant monolayers, was collected for structural
determinations using ^{1}H-nuclear magnetic resonance (NMR) spectroscopy. The resulting spectrum is then compared to the ^{1}H-NMR
spectrum which was obtained beforehand from the partially
purified, microbubble surfactant mixture prior to monolayer
formation and compression.

7.1 ^{1}H-NMR SPECTROSCOPY OF ISOLATED MICROBUB-BLE SURFACTANT

The biochemical isolation of the microbubble surfactant
mixture from forest soil has been described in detail in preceding
chapters. This partially purified surfactant mixture was then
lyophilized and redissolved in CD_3OD. ^{1}H-NMR spectra at 270
MHz were obtained with a Bruker Hx-270 NMR spectrometer
(located at the Australian National University in Canberra) with an
Oxford Instrument Co. superconducting magnet using 4K data
points, pulse length 8 μsec and acquisition time of 0.5 sec. Approximately 1.0% solutions (w/v), in CD_3OD, of the microbubble
surfactant mixture were employed in these measurements, at an
ambient probe-temperature of 26°C. ($CDCl_3$ was also used for
running spectra of model compounds.) Tetramethylsilane ($\delta = 0$)
was used as an internal reference standard (ref. 394).
Fig. 7.1 shows a typical ^{1}H-NMR spectrum obtained with
the partially purified, microbubble surfactant mixture prior to monolayer formation. For comparison, Table 7.1 gives the chemical-
shift data for the proton resonances that can be readily identified in
the ^{1}H-NMR spectra of long-chain acyl lipids (ref. 395-401).

130

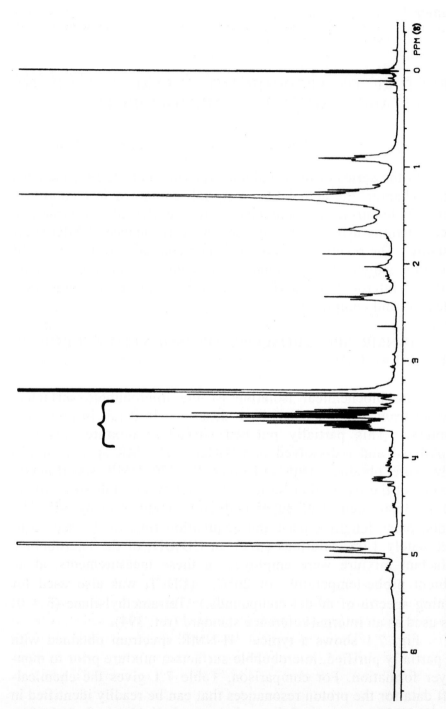

Fig. 7.1. ¹H-NMR spectrum at 270 MHz of the isolated microbubble surfactant mixture. (From ref. 394.)

TABLE 7.1

Chemical shift of protons in long-chain fatty acids and esters (expressed as δ = parts per million relative to tetramethylsilane = 0.00). (Modified from ref. 388.)

δ	Group formula	Description	Type of band
0.9	CH_3	Terminal methyl remote from double bonds and side groups	Unsymm. triplet
1.0	CH_3	Terminal methyl in β-position to olefinic carbon	Triplet
1.3	$CH_2CH_2CH_2$	CH_2 of carbon chain	Broad band
1.6	$CH_3CH=$	CH_3 attached to olefinic carbon	Doublet
1.7	$CH_2-C-C=$	CH_2 in β-position to olefinic carbon	---
2.0	$=CHCH_2$	CH_2 adjacent to isolated olefine group	Apparent doublet
2.25–2.35	CH_2COO	Protons on 2-carbon in acid or ester	Unsymm. triplet
3.7	CH_2-OH	CH_2 of glyceryl mono- or di-ester not esterified in the 1- and/or 3-position	Apparent doublet
3.9–4.1	$CH-OH$	CH of glyceryl ester not esterified in the 2-position	Multiplet
4.2	CH_2-O-C	CH_2 of glycerol esterified in the 1- or 3-position	---
5.1	$CH-O-C$	CH of glycerol esterified in the 2-position	Quintuplet
5.3	$CH=CH$	Isolated double bond protons (not conjugated)	Multiplet

(The actual values depend somewhat on the solvent, concentration, temperature, and conformation of the compound. The values given in Table 7.1 are based on 20% solutions (w/v) in chloroform at 25°C. In general, the observed shifts will lie within $\pm$ 0.1 δ of the figures in the table (see ref. 395).) Accordingly, it can be seen from this comparison that the ^{1}H-NMR spectrum of the partially purified, microbubble surfactant mixture (Fig. 7.1) includes the characteristic peaks for methylene chains (broad peak at δ = 1.3

ppm) and terminal methyl groups (unsymmetrical triplet at $\delta = 0.9$ ppm) of aliphatic lipids. Also present are peaks usually associated with the protons on the 2-carbon in fatty acids and glycerides (unsymmetrical triplet at $\delta = 2.25\text{-}2.35$), the CH_2 of glycerol probably esterified in the 1- or 3-position ($\delta = 4.2$), and CH_3 and CH_2 attached to olefinic carbon ($\delta = 1.6$ and 2.0, respectively). An entirely separate and especially noteworthy feature of the ^{1}H-NMR spectrum of the partially purified, microbubble surfactant mixture is the characteristic family of peaks originating from carbohydrate material (ref. 402,403), specifically the $\underline{H}C\text{–}OH$ protons in sugar residues that occur between δ values of 3.4-3.8 ppm (Fig. 7.1), which will be discussed further in Section 7.3.

7.2 LANGMUIR-TROUGH MEASUREMENTS AND COLLECTION OF MONOLAYERS

For the next phase of this study, monomolecular films were prepared on a subphase of (ultrapure) distilled water. Aliquots, between 50-250 µl, of the ethanol-solubilized microbubble surfactant mixture were applied slowly to the surface of the subphase from a Hamilton microsyringe. It was found unnecessary to allow the films to stand for more than 2 min after spreading before taking measurements. Furthermore, following compression or expansion, the surface pressure was observed to remain constant for periods of up to at least 10 min. All measurements were made at $20.0 \pm 0.5°C$.

For the collection of monolayers, the microbubble surfactant mixture was spread (see above) and compressed to at least 24 dyne/cm. Thereafter, a rectangular piece of stainless steel mesh (4 cm x 1 cm, 30 mesh) was introduced vertically through the surface of the subphase, by means of a small handle, and alternately raised and lowered so that the air/water interface remained within the area of the mesh. This slow, vertical oscillation was coupled with a horizontal movement of the mesh along the surface of the subphase. The procedure resulted in a continual drop in the measured surface pressure (see above) as monolayer material gradually accumulated over the surface of the mesh. The surfactant-coated mesh was then transferred to and shaken in a glass vial containing high-purity ethanol. Once rinsed, the stainless steel mesh was recoated with surfactant film material in the same manner, and rinsed again in the ethanol-filled vial. The process was repeated many times until the measured surface pressure declined to approximately 5 dyne/cm, at which point a fresh monomolecular film was prepared, compressed, and the monolayer collection continued. Numerous spread monolayers were prepared in succession for this purpose.

7.3 ^{1}H-NMR SPECTROSCOPY OF COMPRESSED MONO-LAYER MATERIAL

The collected microbubble-surfactant monolayer material was subsequently lyophilized and redissolved in CD_3OD. As with the preceding NMR measurements (see Section 7.1), ^{1}H-NMR spectra at 270 MHz were obtained with the same Bruker Hx-270 NMR spectrometer. In this case, approximately 0.5% solutions (w/v in CD_3OD) of the microbubble-surfactant monolayer material were employed in the measurements (all taken at 26°C). Tetra-methylsilane was used as an internal reference standard (ref. 394).

Fig. 7.2 shows a typical ^{1}H-NMR spectrum obtained with microbubble-surfactant material collected from compressed mono-layers. Comparison of this spectrum with Fig. 7.1 clearly demonstrates that the characteristic family of peaks originating from carbohydrate material (ref. 402,403), between δ values of 3.4-3.8 ppm, is markedly reduced following monolayer formation and compression. (Note, for example, the relative area and maximum height of the terminal methyl group (δ = 0.9 ppm) versus the carbohydrate family of peaks (δ = 3.4-3.8 ppm) within each of the two spectra (Figs. 7.1 and 7.2).) Contrariwise, the characteristic peaks (cf. Table 7.1) for methylene chains (δ = 1.3 ppm) and terminal methyl groups (δ = 0.9 ppm) of long-chain lipids remain essentially unaltered (Fig. 7.2).

However, the peaks usually associated (cf. Table 7.1) with CH_3 and CH_2 attached to olefinic carbon (δ = 1.6 and 2.0, respectively) show a relative reduction in height and area. This finding suggests that compression of the microbubble-surfactant monolayer results in the ejection of some unsaturated lipids, as well as most of the carbohydrate material, from the monolayer. Such a conclusion is consistent with the frequently mentioned

134

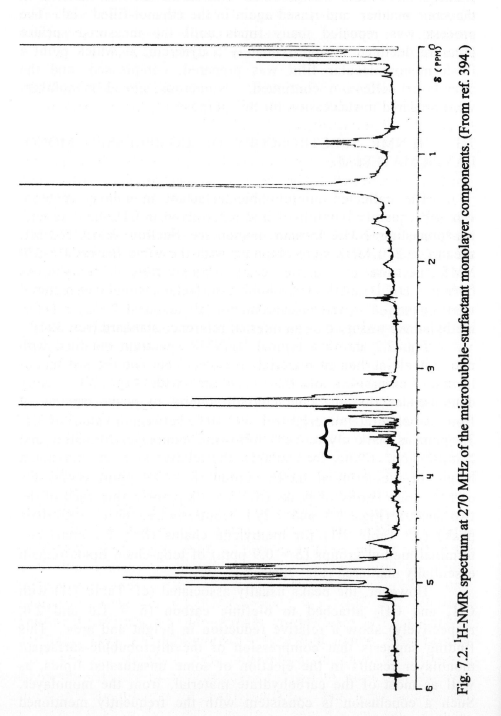

Fig. 7.2. ^{1}H-NMR spectrum at 270 MHz of the microbubble-surfactant monolayer components. (From ref. 394.)

finding, of various investigators (e.g., ref. 17,115-118) cited in Chapter 1, that <u>saturated</u> acyl lipids are the most effective surfactants in reducing gaseous diffusion across compressed monolayers at the gas/water interface (see Sections 1.2.2 and 1.3.1).

7.4 CHEMICAL SIMILARITIES BETWEEN MICROBUBBLE-SURFACTANT MONOLAYERS AND LIPID SURFACE FILMS AT THE AIR/SEA INTERFACE

Several studies by various investigators of lipid surface films at the air/sea interface provide interesting clues as to likely structural candidates for the suspected saturated acyl lipids which primarily surround the long-lived microbubbles in natural waters. More than three decades ago, Garrett (ref. 86,404) concluded that the materials selectively adsorbed at the sea surface are mostly acyl lipids, because of their especially high surface activity. As Johnson and Cooke later point out (ref. 41), this selectivity demonstrated by Garrett for surface films should be even more in evidence for rising, partially dissolving microbubbles because the microbubbles pass through sea water and contact more surface-active material than does a static surface film. The surface area of such a microbubble also contracts (in the process of forming a film-stabilized microbubble) and thus provides a dynamic impetus for the expulsion of material of lower surface activity (ref. 41). A large fraction of these microbubbles, thereafter, carry their surfactant coating to the sea surface (ref. 85).

The lipid surface film consists, under natural conditions, mainly of saturated and unsaturated glycerides and fatty acids produced in the aquatic environment (ref. 405,406). Multifilms appear to form as a consequence of saturation (ref. 407). Interestingly, the amount of neutral acyl lipids (i.e., glycerides) relative to fatty acids "is always higher in the surface multifilms than in the subsurface water" (ref. 405,406). This environmental chemical finding is quite consistent with the earlier-described physicochemical results from microbubble experiments, involving agarose gels, which indicated that microbubbles were stabilized primarily by nonionic surfactants (see Sections 2.2 and 2.3). Another reassuring finding from sea-surface-film studies concerns the results from gas chromatograms of unpolluted sea water samples: the relative amount of saturated acyl lipids is higher in

136

the surface film compared to the subsurface water (ref. 405). This oceanographic finding, too, agrees well with previously described microbubble-related studies which indicated that saturated acyl lipids are the most effective surfactants in stabilizing microbubbles (see Sections 1.2.2, 1.3.1, and 7.3). Taken together, these two chemical oceanographic findings suggest that the most likely structural candidates for the predominant acyl lipids stabilizing microbubbles are <u>saturated glycerides</u>.

Chapter 8

STABLE MICROBUBBLES IN PHYSIOLOGICAL FLUIDS: COMPETING HYPOTHESES

In the preceding chapter, evidence was presented that indicates saturated glycerides are the predominant surfactants surrounding long-lived gas microbubbles in natural waters. Saturated glycerides are, at the same time, widely known to be common biochemical constituents in the blood and tissues of mammals and other animals (ref. 270). These facts, taken together, bring to mind the question of whether stable microbubbles also occur naturally in biological fluids.

The possibility of significant concentrations of surfactant-stabilized gas microbubbles existing in physiological fluids, and which act as bubble nuclei, has been used for several years as a plausible explanation for the etiology of decompression sickness in animals (ref. 29,49,55,56,114,135,138,139,147). As concerns humans in particular, decompression sickness most often occurs from underwater diving (ref. 29,49,56,114,135), caisson operations (ref. 55), or other compressed-air work. The disease, which involves macroscopic bubble formation and growth in body fluids and tissues, is associated with too rapid a return by an individual to atmospheric pressure after exposure to a hyperbaric breathing environment. The clinically safe, maximum rate of return to atmospheric pressure will depend upon both the amplitude of the hyperbaric exposure (e.g., water depth) and its duration; moreover, for any given hyperbaric exposure, a wide variety of decompression schedules are in use for humans in an effort to prevent the disease (e.g., ref. 56). Interestingly, the relative effectiveness of these different decompression tables in preventing decompression sickness in field operations has been found (see Section 8.1) to depend upon two major design factors, both of which are shown below to also govern bubble formation in agarose gels -- where the bubbles specifically arise from surfactant-stabilized microbubbles (cf. Chapters 2 and 3). This correlation is of interest since it seems to suggest that surfactant-stabilized gas microbubbles may in fact exist in physiological fluids also.

8.1 COMPARISON OF DIFFERENT DECOMPRESSION SCHEDULES: CORRELATION BETWEEN BUBBLE PRODUCTION IN AGAROSE GELS AND INCIDENCE OF DECOMPRESSION SICKNESS

8.1.1 Background observations

The U.S. Navy Standard Air Decompression Tables (USN) have been, and continue to be, used extensively to guide human decompression procedures throughout much of the world. These tables (ref. 408), as well as the French Navy decompression table (FN) and the Japanese Department of Labor standard decompression table (Japan), are all based on the Haldane-ratio principle (ref. 409,410). However, other tables have been developed that do not employ the Haldane-ratio principle to guide decompression, and several of these tables appear to be considerably safer (ref. 411-413).

The Haldane-ratio principle (ref. 409) states that it is the ratio of (tissue gas tension)/(ambient pressure) = p_{tis}/p_{amb} , and not the difference (or supersaturation pressure) $p_{ss} = p_{tis} - p_{amb}$, that determines how rapidly a diver can surface safely. To compare these points of view, some investigators (ref. 135,414) have subjected mammalian gelatin samples to simple pressure schedules for which p_{tis}/p_{amb} was varied while $p_{ss} = p_{tis} - p_{amb}$ was held constant. Although the difference in the Haldane ratios were large, very little difference was found in the mean number of bubbles formed in the gelatin samples. Accordingly, Yount et al. proposed a more effective decompression procedure that is not based on the Haldane-ratio principle (ref. 135,414).

The Royal Naval Physiological Laboratory (RNPL) has also found reason to doubt the Haldane concept of bubble formation and has developed its own decompression schedule (ref. 411). The RNPL table and the decompression table used by the French Ministry of Labor (FL) are based on diffusion theory (ref. 411). The most characteristic feature of these tables is the deeper first stop during decompression after diving, with the result that the total decompression time is prolonged to approximately 3 times the period specified by the Haldane-ratio-principle decompression schedules for the same dive (see below). Similarly, a subsequently developed decompression table (Model 1), used by Nippon Salvage Company of Tokyo in underwater operations at Guam (ref. 415), employs both a deeper first stop and longer total decompression time than do the corresponding USN tables. By using the Model 1 table instead of the USN tables, Nippon Salvage Company was able to reduce the incidence of bends among its divers from over 1.5% to only 0.35% (ref. 415,416).

In a related experimental study, the relative effectiveness of all the above-mentioned decompression tables in reducing bubble formation within aqueous gels was evaluated quantitatively under rigorously controlled conditions; specifically, visual counts were conducted of the bubbles formed in highly purified agarose gels (ref. 49,139) subjected to the different decompression schedules.

8.1.2 Methods

The pressure vessel, glass counting chambers, and their plexiglass holder used in these experiments are all similar to those described earlier (ref. 139). As viewed through the plexiglass wall of the pressure vessel, the inside dimensions of each rectangular counting chamber are 15 mm wide and 6 mm along the line of sight.

Each chamber was filled with agarose solution to a depth of 4 mm. After gelation, the agarose samples were exposed to 100-fsw (feet sea water) pressures (i.e., 44.5 psig) for 40 min at 21°C, and then decompressed to atmospheric pressure in accord with one of the seven different decompression schedules tested (see Section 8.1.1). Only bubbles formed in the bottom 3 mm of a given agarose sample were counted, so that the total volume of gel examined in each sample amounted to 0.27 ml.

To form the agarose gel itself, a stock solution containing 1.0 mM HEPES (N-2-hydroxyethylpiperazine-N'-2-ethanesulphonic acid, pH_a 7.55) buffer was first prepared. The pH was adjusted to 7.4 by adding small amounts of NaOH. After heating an aliquot of the buffered solution to 80°C, highly purified agarose powder from Bio-Rad, lot no. 14672 (ash < 0.5%, sulfur content < 0.1%), was added (1% w/w) to the aliquot and dissolved by gently swirling the solution. After 10 min, the agarose solution was transferred to a 50°C bath, from which it was pipetted into the acid-cleaned counting chambers.

8.1.3 Experimental results

The actual depth profile of each of the seven decompression schedules tested in this study is shown in Fig. 8.1. For the standard dive chosen of 100-fsw depth and 40-min duration, it can be seen from the figure that there is a wide variation in the decompression profiles specified by the different tables examined. (The source of each decompression schedule is given at the upper right of the figure, along with total decompression time (min) in parentheses. Data taken from ref. 56, copyright 1979 Undersea Medical Society, Inc.) For example, the total decompression times required by the different military and commercial decompression tables ranged between 12.00 min (FN table) and 58.24 min (RNPL table). The shortest total decompression time of only 11.04 min was required by the schedule of Yount et al., which is not a commercial table but rather a pressure schedule derived from work with mammalian gelatin (ref. 135,414).

Eight trials using agarose gels were performed for each of the seven decompression schedules, and the total number of bubbles formed per gel sample (0.27 ml) was recorded. Table 8.1 lists the number of bubbles (mean ± S.E.M.) observed with each decompression profile at various depths, i.e., "stops".

8.1.4 Water depth at first stop, and total decompression time

The separate decompression tables of the French Navy, the U.S. Navy, and the Japanese Department of Labor, which are all based on the Haldane-ratio principle (ref. 408-410), require total decompression times for the test dive which are much shorter than those required by other military and commercial tables (Table 8.1). The first stop during decompression with either the FN, USN, or Japan tables occurred at a 10-ft depth, and the mean bubble counts (± S.E.M.) within the 0.27-ml agarose samples just prior to termination of this first (and only) stop were 127.25 ± 9.39, 111.88 ± 17.64, and 98.75 ± 10.72, respectively. Of these three Haldane-ratio-principle tables, the FN table required the shortest total decompression time and the Japan table the longest time, so that the mean bubble number at the 10-ft depth was inversely related to the total decompression time. (In these three cases, the total decompression time essentially represented the sum of the initial

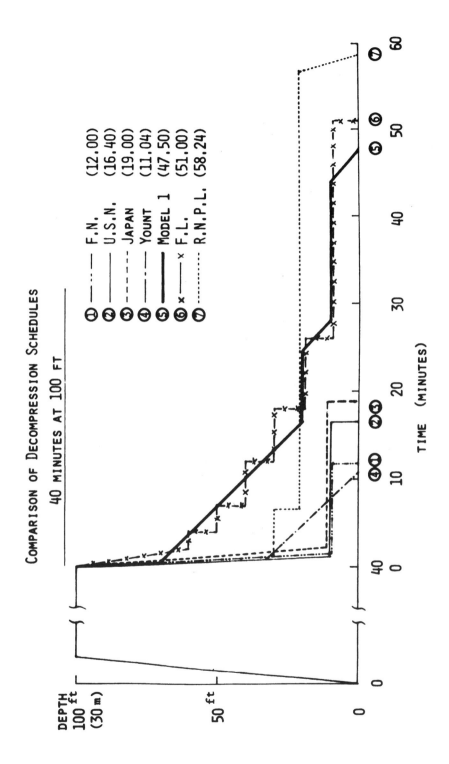

Fig. 8.1. Comparison of seven decompression profiles for standard dive of 100-fsw depth and 40-min long.

142

TABLE 8.1

Bubble formation in agarose gels subjected to various decompression schedules[a]. (Slightly modified from ref. 56, copyright 1979 Undersea Medical Society, Inc.)

Schedule	Total time, min	Number of bubbles, mean of 8 trials $\pm$ S.E.M.		
		20-ft stop	10-ft stop	Final
French Navy	12.00		127.25 $\pm$ 9.39	129.88 $\pm$ 9.59
U.S. Navy	16.40		111.88 $\pm$ 17.6	115.38 $\pm$ 18.2
Japan, Dept. of Labor	19.00		98.75 $\pm$ 10.7	109.00 $\pm$ 6.90
Yount et al.	11.04			98.63 $\pm$ 5.91
Model 1	47.50	25.75 $\pm$ 3.83	54.87 $\pm$ 2.24	61.13 $\pm$ 1.12
France, Dept. of Labor	51.00	22.25 $\pm$ 4.90	74.13 $\pm$ 6.96	74.87 $\pm$ 5.73
RNPL	58.40	28.13 $\pm$ 7.16		72.38 $\pm$ 3.58

[a]Following a "bottom time" of 40 min with air at 44.5 psig ($\approx$ 100 fsw); agarose samples were 1% w/w, 0.27 ml in volume.

period of ascent plus the time spent at the single decompression stop, with both these parameters varying across cases.) The final bubble number (Table 8.1) was also inversely related to the total decompression time: 129.88 $\pm$ 9.59 for the FN table (12.00 min), 115.38 $\pm$ 18.28 for the USN table (16.40 min), and 109.00 $\pm$ 6.90 for the Japan table (19.00 min).

The shortest total decompression time specified by any of the schedules (11.04 min) was that of the Yount et al. schedule (ref. 135,414) for the standard dive. Nonetheless, the final number of bubbles produced by this particular schedule, 98.63 $\pm$ 5.91 per agarose sample (0.27 ml), was less than that produced by any of the three tables (FN, USN, Japan) based on the Haldane-ratio principle (Table 8.1). In particular, the final bubble count produced by the FN schedule was markedly higher ($P < 0.005$), despite the fact that this schedule involved almost the same total decompression time, i.e., 12.00 min. The Yount et al. schedule has no stop during decompression, but instead initiates a slow and constant rate of decompression at about a 40-ft depth (ref. 135.414). The starting point for the slow decompression is, therefore, approximately 3.3 times deeper than with the tables (FN, USN, Japan) based on the

Haldane-ratio principle (see Fig. 8.1).

As with the Yount et al. schedule, the Model 1, FL, and RNPL tables are based, at least in part, on diffusion theory (ref. 411). Accordingly, these latter three tables also require that slow decompression commence at a greater depth than that specified by the tables based on the Haldane-ratio principle (FN, USN, Japan) (see Fig. 8.1). The Model 1 schedule initiated slow decompression at the greatest depth (70 ft (Fig. 8.1)), with the result that the mean number of bubbles produced per 0.27-ml agarose sample (Table 8.1) was significantly less than the number produced by the FL schedule at the 10-ft stop (54.87 ± 2.24 versus 74.13 ± 6.96, respectively), and was significantly less than the numbers produced by either the FL schedule or the RNPL schedule at the end of decompression (61.13 ± 1.12 versus 74.87 ± 5.73 and 72.38 ± 3.58, respectively) ($P < 0.05$). This significantly reduced production of bubbles by the Model 1 schedule occurred despite the fact that the total decompression times of the FL and RNPL schedules were longer (47.50 min versus 51.00 min and 58.24 min, respectively). Therefore, as already indicated above by the agarose results obtained from use of the Yount et al. schedule, it is clear that the depth at which slow decompression commences is a major factor, along with total decompression time, in determining the extent of bubble formation. Accordingly, it is the greater magnitude of both of these contributing factors that explains the far fewer bubbles produced in agarose gels by the Model 1, FL, and RNPL tables, as compared with the FN, USN, and Japan tables ($P < 0.01$).

8.2 COMPARISON OF CAVITATION THRESHOLDS FOR AGAROSE GELS AND VERTEBRATE TISSUES

Apart from the above-described correlation between bubble production in agarose gels and the incidence of decompression sickness in humans, there exists cavitation-threshold data from simple vertebrates which also indirectly suggests that surfactant-stabilized gas microbubbles exist in physiological fluids. As pointed out in Section 2.1, the actual threshold for bubble formation in agarose gel, upon rapid decompression from either hyperbaric or atmospheric pressure, is between -3 and -4 psig ($\approx$ -0.25 atm) for either nitrogen, helium, or carbon dioxide gas (see Figs. 2.1 and 2.2). These results are in excellent agreement with

144

earlier findings of substantial bubble production in fish (ref. 417) with a gas tension difference of -4.5 psig between the tissue and the ambient hydrostatic pressure of water, while the actual cavitation threshold measured in bullfrogs (ref. 418) is -3 psig.

8.3   CONTRADICTORY FINDINGS

While confirming many of the above findings on cavitation thresholds in vertebrate tissues, a later study by McDonough and Hemmingsen (ref. 419) reviews much data which indicates that species can vary widely in their susceptibility to bubble formation. For example, these authors point out that unicellular micro-organisms and primitive multicellular animals can be subjected to very high gas supersaturations without bubbles forming in them (ref. 424-426). In contrast, higher animals are quite susceptible to bubble formation. Humans will develop decompression sickness (i.e., the "bends") when exposed to air supersaturation of 1 atm (ref. 427). Fish are even more susceptible; some species contract "gas bubble disease" when kept in water supersaturated by just 0.2 atm (ref. 428).

McDonough and Hemmingsen (ref. 419) confirm that for bubbles to develop in vertebrates from such low gas super-saturations, some mechanism or structure must promote the initial in vivo bubble nucleations. They cite, as one initial possibility, the popular, general hypothesis that animals contain a reservoir of microscopic gaseous nuclei in the body fluids or tissues, which expand into bubbles when the organism is decompressed (ref. 2). These authors point out that results consistent with this hypothesis have been obtained with shrimp (ref. 429) and rats (ref. 430), where the application of relatively high hydrostatic pressure before decompression apparently reduced the incidence of bubble formation, presumably by forcing potential gas nuclei into solution before they could serve as bubble precursors (ref. 419).

However, findings that do not support this mechanism include the observations by Harvey et al. (ref. 431) that bubbles do not form in mammalian blood or isolated frog tissues decom-pressed to 0.031 atm. Steelhead trout fingerlings also are not affected by exposure to 0.16 atm (ref. 432). McDonough and Hemmingsen (ref. 419) argue that in these experiments, the reduction in external pressure should have caused the gas nuclei present to expand to a detectable size.

These authors also explain that another possible mechanism for bubble formation, proposed several times in the past, is by tribonucleation at points of rubbing contact between solid structures immersed in mildly gas-supersaturated solutions (ref. 433,434). Viscous adhesion between the moving surfaces has been proposed to develop tension in the liquid sufficient to create cavities leading to bubble formation (ref. 435). The super-saturations required for this mechanism are inversely related to the speed of movement and viscosity of the liquid (ref. 436). McDonough and Hemmingsen further point out that Ikels (ref. 436) suggested that in vivo tribonucleation might occur between articulating surfaces of joints, at muscle tendon inserts, or in small blood vessels subjected to collapse and expansion. Similarly, it has been proposed that bubbles may originate in physiological fluids through tensions produced in the body fluids by muscular contraction (ref. 2,418).

In view of these conflicting data and competing hypotheses, McDonough and Hemmingsen (ref. 419) performed a detailed study with crustaceans to evaluate the occurrence of the proposed bubble formation mechanisms in these animals -- since they are similar to vertebrates in possessing extracellular fluid, lipid storage, and muscular systems, yet are relatively transparent and allow bubbles formed by decompression to be seen without dissection. A procedure of compression/gas equilibration/decompression was used to determine the argon and helium supersaturation levels necessary for bubble formation in a variety of crustaceans. In their study, these authors defined: "supersaturation threshold" as the gauge equilibration pressure necessary to produce bubbles in at least 50% of the animals when decompressed to ambient atmospheric pressure; "gut bubbles" as bubbles that were com-pletely confined to the lumen of the gut; and "somatic bubbles" as bubbles found in the true body fluids or tissues of these animals (ref. 419).

It was found from their experiments (ref. 419) that pre-pressurization generally had little effect on somatic bubble formation. The results with shore crab larvae were especially striking considering the low resistance of these animals to bubble formation and the large difference between the hydrostatic and gas equilibration pressures used. Furthermore, when a slow compression schedule was used with adult sand crabs, the number of animals with somatic bubbles did not increase. McDonough and Hemmingsen emphasize that studies on gelatin model systems (ref. 135) show that prepressurizations and slow compressions can drastically alter the formation of bubbles from gas nuclei; 75% of the bubbles forming from nuclei in gelatin supersaturated by 10 atm air could be prevented by just a 1-sec prepressurization to 20 atm, whereas slow compressions increased the number of bubbles by a factor of 10 (ref. 419).

Other observations were also inconsistent with the hypothesis that gas nuclei may play an important role in the formation of somatic bubbles. For brine shrimp, copepods, and shrimp larvae, the helium supersaturation thresholds were nearly twice as high as those for argon. McDonough and Hemmingsen (ref. 419) argue that such a large difference would be expected for homogeneous nucleation (ref. 190), but not if the bubbles were forming from gas nuclei. They cite, as an example, past experiments with agarose gels (ref. 139) which demonstrated that carbon dioxide, nitrogen, and helium produce approximately equal numbers of bubbles in supersaturated nuclei-containing gel, even though these gases differ greatly in solubility (see Fig. 2.1).

The gut bubbles in adult brine shrimp did appear, however, to form from gas nuclei (ref. 419); these presumably were incidentally ingested by the animals during filter feeding. Thus a slow compression schedule increased the number of bubbles by apparently preserving nuclei during compression to the equilibration pressure. At each pressure level, gas could diffuse into the gas nuclei, tending to stabilize them against collapse when further compressed. Prepressurization had the opposite effect, as it would tend to reduce the number of bubbles as a result of presumed dissolution of many gas nuclei (ref. 419; see also Sections 1.3.1 and 1.4.3).

In summary, McDonough and Hemmingsen (ref. 419) conclude that the extreme resistance of the brine shrimp larva to bubble formation is consistent with the hypothesis that the intracellular environment of eucaryotic cells is intrinsically very resistant to bubble nucleation. The added results for the adult brine shrimp, copepods, and larval decapods show that resistance can still be quite high, even though circulatory systems and lipid storage depots are present. Finally, the much lower resistance of the shore crab larvae and adult decapods is not due to the existence of preformed gas nuclei in the body fluids or tissues. Instead, McDonough and Hemmingsen believe that bubbles may be induced in the hemolymph by tribonucleation or by other mechanisms that cause mechanical stress at limb joints. For this process to occur, limb movements must take place after the decompression but before the gas supersaturation has dropped by diffusion below the threshold level (ref. 419; see also ref. 420-423).

In support of their position favoring tribonucleation in crustaceans, McDonough and Hemmingsen (ref. 419) summarize results obtained with shore crab larvae which provide evidence for bubbles induced by movement. The number of bubbles that formed in these animals decreased with decreasing degree of limb movement, such as was produced by hypoxia or by high pressure per se, which is known to have an immobilizing effect on crustaceans through its action on the nervous system (ref. 437). Conversely, the increase in movements in crab larvae exposed to formaldehyde greatly increased bubble formation. (It was considered unlikely that quantities of formaldehyde sufficient to directly influence bubble formation in the body fluids could have entered these animals in the brief period before bubbles appeared. The location of bubble nucleation was also the same for these formaldehyde-exposed (and decompressed) animals as for untreated (but decompressed) crab larvae (ref. 419).)

8.4   HOMOGENEOUS NUCLEATION HYPOTHESIS

Although the preceding section provides much evidence to indicate that tribonucleation, rather than preformed gas nuclei, is the major causative factor for bubble formation in the fluids and tissues of crustaceans and possibly also vertebrates, there exists at least one other plausible explanation for the high susceptibility of vertebrates to bubble formation upon decompression. As Weathersby et al. explain (ref. 438), several current theories of decompression sickness presume the pre-existence of gas bubble nuclei in tissue because the de novo nucleation (i.e., homogeneous nucleation) of gas bubbles in the body is thought to be theoretically impossible. Re-examination of nucleation theory reveals the over-whelming importance of two parameters: gas supersaturation and tissue surface tension. For the high surface tension of pure water, homogeneous nucleation theoretically requires more than 1,000 atm supersaturation. However, lower values of surface tension allow homogeneous nucleation to occur with vastly smaller super-saturations. Furthermore, application of homogeneous nucleation theory can provide reasonable fits to both rat and human pressure-reduction data with values of surface tension within a range reported for some biological fluids (i.e., below 5 dynes/cm), e.g., lung surfactant (ref. 438).

8.5   CLINICAL USE OF INJECTED GAS MICROBUBBLES: ECHOCARDIOGRAPHY; POTENTIAL FOR CANCER DETECTION

Consistent with the more recent evidence indicating that stable microbubbles might not occur naturally in physiological fluids (see Sections 8.3 and 8.4), including those of vertebrates, are the findings from ultrasound measurements within the mammalian cardiovascular system. Gross et al. (ref. 439) used a resonant bubble detector system to monitor for gas microbubbles approximately 4 μm in diameter (i.e., resonant in the low megahertz frequency range) within the circulating blood in vivo. They were unable to find gas microbubbles in 80 separate 2 to 5 min observations made in 23 different subjects representing 3 separate species. Twenty-five in vivo experiments were conducted during which the subject's cardiovascular system was exposed to continuous wave ultrasound of 1 MHz and/or 1.6 MHz and spatial peak intensities ranging from 125 mW/cm^2 to 32 W/cm^2. No evidence of ultrasonic cavitation within the cardiovascular system was found in any of these experiments. In a separate group of control experiments, these authors were able to demonstrate that the addition of cavitation nuclei to the cardiovascular system (while it was being irradiated with ultrasound at intensities > 4 bar) did, in fact, result in entrapment of the bubbles within the acoustic beam (ref. 439).

This last-mentioned successful series of control experiments, involving addition of cavitation nuclei to the cardiovascular system, brings to mind at least one feasible medical application for injected (synthetic) surfactant-stabilized microbubbles (cf. Chapters 9 and 10) as concerns evaluation of cardiovascular function. Various types of size-controlled, nontoxic, synthetic microbubble preparations have found clinical application, for more than a decade, as standardized echo contrast agents in echocardiography (for reviews see ref. 445,446).

Earlier studies (ref. 440-442) with ordinary air microbubbles (without any synthetic surfactant coating) have already shown that echocardiographic contrast produced by microbubbles is useful in the qualitative analysis of blood flow and valvular regurgitation. In addition, quantitative studies (ref. 440) have shown a correlation between individual contrast trajectories on M-mode echocardiography and invasive velocity measurements in human beings. Meltzer et al. (ref. 441) have shown that velocities derived from the slopes of contrast trajectories seen on M-mode echocardiography correlate with simultaneous velocities obtained by Doppler techniques. (This correlation is expected because both measures represent the same projection of the microbubble velocity vector, that is, in the direction of the sound beam.) More detailed studies (ref. 442) confirmed that microbubble velocity obtained from either Doppler echocardiography or M-mode contrast trajectory slope analysis correlates well with actual (Doppler-measured) red blood cell velocity. Thus, these early studies have shown that microbubbles travel with intracardiac velocities similar to those of red blood cells.

As noted in these early clinical studies using ordinary air microbubbles (e.g., ref. 442), there is a significant potential for variability in microbubble size based on subtle variations in injection technique, and perhaps the distance between the injection site and the heart. The potential introduction of larger microbubbles might provide a different spectrum of velocities. Because the size spectrum of microbubbles actually imaged in the heart is unknown, the significance of this potential source of variability would be unclear. All of the above-mentioned sources of variability can be greatly reduced or eliminated by the use of certain synthetic microbubbles which, as will be described in Chapters 9 and 10, are very long-lived (because of their particular surfactant coating), of controlled size, and nontoxic.

Apart from echocardiography, another promising clinical application of synthetic microbubbles is the ultrasonic monitoring of local blood flow in the abdomen (analogous to the earlier use of gas microbubbles to monitor myocardial perfusion (ref. 443)). Such refined ultrasonic blood flow measurements, utilizing injected synthetic microbubbles, can provide better clinical detection of tumor neovascularization as well as any subtle changes in the normal vascularization patterns of organs neighboring abdominal masses (cf. ref. 445,446). Hence, through the use of synthetic microbubbles, ultrasound can now provide earlier diagnosis of many abdominal masses; such early detection may well improve treatment of several classes of serious abdominal cancers, a notorious example being liver tumors.

In actuality, over the last two decades many types of synthetic-microbubble contrast agents, each stabilized by a different chemical coating, have been developed by different research groups and/or companies for use in contrast-assisted echocardiographic, myocardial perfusion, and/or tumor detection studies (for literature reviews see ref. 444-446). However, only the particular coated-microbubble agent described in Chapters 9-11, which was modeled primarily from natural microbubble surfactant, contains specifically nonionic lipids exclusively throughout the microbubble coating; interestingly, this nonionic-lipid-monolayer coating causes this particular synthetic-microbubble agent to display surprising tumor-targeting abilities useful for both diagnosis and treatment of tumors (see Chapters 12-14).

Chapter 9

CONCENTRATED GAS-IN-LIQUID EMULSIONS IN ARTI-
FICIAL MEDIA. I. DEMONSTRATION BY LASER-LIGHT
SCATTERING

9.1 PHYSIOLOGICAL HINTS FOR THE PRODUCTION OF
ARTIFICIAL MICROBUBBLES

The overall impression one obtains from the various studies
and competing hypotheses described in Chapter 8 is that stable gas
microbubbles more likely do not occur naturally in physiological
fluids; on the other hand, injected artificial surfactant-stabilized
microbubbles are likely to have useful clinical applications.
Interestingly, these proposed long-lived, artificial microbubbles
would resemble in many ways (including their having a low-
dielectric material in their core) the biochemical assemblies, i.e.,
oil-in-water emulsions, already formed in the human body during
the process of digestion and absorption of fats from inside the
small intestine. For example, once digested fat is absorbed into the
intestinal epithelium, the fat is processed further and organized into
protein-coated lipid droplets called "chylomicrons" (ref. 447-449).
The 0.1-3.5 μm diameter sizes of these structures (ref. 270,448)
closely resemble the size distribution, usually < 5 μm, of long-lived
gas microbubbles (see Sections 3.2 and 4.1.1). Moreover, the
protein content of chylomicrons even though low (about 2% by
weight) is a very consistent feature (ref. 447-450), as is also true
for natural microbubble surfactant (see Chapters 3-5) where the
protein content is less than 5% by weight (see Chapter 6). The
protein coat of chylomicrons, even though incomplete and mixed
with various lipids in the interfacial film, helps keep them
suspended and prevents them from sticking to each other or to the
walls of the lymphatics or blood vessels (ref. 449,450). (Similarly,
one might expect that the protein content of natural microbubble
surfactant (see Section 1.3.1 and Chapters 3-5) helps prevent the

coalescence and/or destruction of the stable gas microbubbles known to be present in natural waters (see Chapters 1, 3, 4, and 5).)

However, the presence of a partial protein coat (ref. 194,451,452) is not, of course, an indispensable requirement for all (or even most) stable, spontaneously forming oil-in-water emulsions (ref. 194,453-463).   Returning to another related physiological example from the intestinal processing of fats in humans, undigested fat enters the intestine in the form of large lipid globules which undergo emulsification through the detergent-like action of various cholesterol derivatives, i.e., the bile salts (ref. 447-449).   This emulsification facilitates subsequent enzymatic degradation (by pancreatic lipase) of the triglycerides to monoglycerides.

The monoglycerides, cholesterol derivatives (bile salts), cholesterol, and other lipid components then spontaneously form stable "mixed micelles" (ref. 447,448) in the intestinal lumen. Parallel consideration of results from related artificial oil-in-water emulsion studies offer the suggestion, however, that if the monoglycerides had sufficiently long chain lengths and the rest of the interfacial film contained only cholesterol and/or cholesterol derivatives, one could expect larger, spontaneously forming structures (containing low-dielectric material in their core) to be formed.   For example, Schulman et al. (ref. 454) reported that "it has been shown that the chain length of a normal alcohol, when benzene was the oil phase and potassium oleate the micelle forming compound, was a governing factor in controlling the size of the droplets.   Increasing the chain length greatly diminished the range of the phase diagram such that above decyl alcohol, no microemulsions could be formed with this system.   Only coarse emulsions could be formed of 0.5 μm diameter droplets.   This occurred also with systems stabilized with cetyl alcohol or cholesterol … and thus, although the emulsions were spontaneous-ly formed, no microemulsions could be made.   It was considered that the interfacial mixed monolayer was too highly condensed with the strong associating components, and thus no great degree of curvature necessary for the very small droplets could be produced, a vapor condensed film being essential" (ref. 454).

Accordingly, a detailed series of tests in this laboratory revealed that aqueous solutions of saturated monoglycerides (with acyl chain lengths greater than 10 carbons) combined with cholesterol and cholesterol derivatives did, in fact, form stable <u>gas-in-water</u> emulsions when shaken vigorously (in an air atmosphere). Evidence for the formation of these significantly stable (hours to days) emulsions is presented below.

9.2 LASER-BASED FLOW CYTOMETRY AND FORWARD-ANGLE LIGHT SCATTERING

All experimental measurements were carried out with a Coulter EPICS V System. This instrument is a laser-based flow cytometer which, as one of its simpler analytical functions, utilizes light scatter measurements to accurately size cells or similar particles (e.g., artificial surfactant-stabilized microbubbles) suspended in aqueous media. The light scatter measurements are sensitive to particle sizes as small as 0.3 µm in diameter.

To make the measurements, particles in liquid suspension are presented under mild pressure to a flow cell where they are surrounded by a laminar sheath of particle-free liquid. This coaxial stream then exits through a flow chamber as a (76-µm diameter) jet. This hydrodynamic focusing, which is ordinarily bubble-free when only solid particles are measured, insures that all particles follow the same path through the detection zone. Sample pressure, sheath pressure, and particle concentration are controlled so that particles are presented one-at-a-time to the beam of a UV-enhanced argon-ion laser. As the particles pass through the laser beam, the scatter emissions are collected by detectors which convert the signals to voltage pulses proportional to the amount of light scattered. The pulses are amplified, converted to digital form, and displayed in numeric or histogram formats. In this study, light scattered in specifically the forward direction was monitored, since it is generally proportional to the cross-sectional area (radius^2) of large-size particles (i.e., on the order of a micron) (ref. 129).

9.3 SYNTHETIC MICROBUBBLE COUNTS VERSUS THE CONTROL

A mixture of saturated monoglycerides (with acyl chain lengths greater than 10 carbons) combined with cholesterol and (nonionic) cholesterol derivatives, all initially in powdered form, was used to form high concentrations of artificial gas microbubbles

154

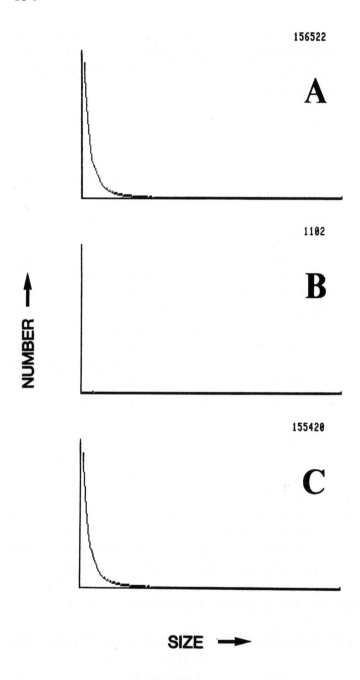

156522

A

1102

B

155420

C

NUMBER →

SIZE →

Fig. 9.1. Synthetic microbubble histograms determined by laser-light scattering in aqueous media (at 21°C): (A) saturated solution of Filmix 3 surfactant mixture; (B) distilled water alone; (C) computed difference in above two histograms. (See text for further discussion.)

in distilled water. This proprietary surfactant mixture, CAV-CON Filmix 3, contains approximately equal parts of monoglycerides and sterol products. A saturated aqueous solution of the surfactant mixture was simply shaken vigorously for a few seconds, in an air atmosphere at room temperature, to form the concentrated gas-in-water emulsions.

Evidence for the formation of synthetic microbubbles was obtained by comparing the particle histograms (Fig. 9.1) for identically preshaken, equivolumetric samples of the artificial-microbubble-surfactant solution (Fig. 9.1(A)) versus distilled water alone (Fig. 9.1(B)). Collection of the scatter emissions, for a total period of 400 sec, commenced one minute after shaking ended in both cases; this one-minute delay allowed essentially all macroscopic bubbles formed in both samples to rise to the surface. It can be seen from the histograms (of identical scale), shown in Fig. 9.1, that the amount of microscopic particles remaining over several minutes in the artificial-microbubble-surfactant solution (Fig. 9.1(A)) was enormously greater than in an equal volume of the distilled water alone (Fig. 9.1(B)); this finding is further confirmed by the total particle counts (over the 400-sec collection period) shown in the upper right corner of both Figs. 9.1(A) and 9.1(B), which differed by more than two orders of magnitude, as well as by the computed difference in the two histograms shown in Fig. 9.1(C).

Additional evidence that the large number of microscopic particles detected in Fig. 9.1(A) are, in fact, artificial surfactant-stabilized microbubbles includes the following: 1) Prior to shaking and microbubble formation, the surfactant solution was passed through a membrane filter which removed all solid debris greater than 6.0 μm in diameter; 2) The detection limit of the instrument, i.e., particle sizes down to 0.3 μm in diameter, strongly argues against the particle count being a representation of a micelle population (which will also be present, but undetected in these experiments). This is especially true since forward-angle light scattering was used, which favors the detection of larger particles in the detection range (0.3-40 μm) of the instrument; 3) Since none of the surfactants used are liquids, oil-in-water microemulsions/emulsions could not be the basis of the particle histogram obtained; 4) In accompanying decompression tests, it was found, in the course of developing Filmix 3, that those surfactant solutions which produced the higher concentrations of growing bubbles upon decompression (i.e., below 1 atm) similarly produced the greater degree of light scatter in the absence of decompression.

An estimate of the actual concentration of synthetic micro-bubbles present in the (shaken) artificial-microbubble-surfactant solution, represented by Fig. 9.1(A), is given by the fact that 360-400 particles/sec were consistently detected at a flow rate of approximately 1 ml/30 min in order to produce the histogram shown. Therefore, the calculated approximate concentration of synthetic microbubbles in the sample is 7×10^5 microbubbles/ml. (A similar calculation for the distilled water sample shown in Fig. 9.1(B) results in an estimated concentration of only 5×10^3 microbubbles/ml.)

9.4   MICROBUBBLE FLOTATION WITH TIME

In the next experiment, the relative size distribution of the artificial microbubbles formed with the Filmix 3 surfactant mixture (described in the preceding section) was monitored over a 1,000-sec period (Fig. 9.2). While the Coulter EPICS V System used did not record the absolute sizes of microbubbles detected, relative sizes were indicated by the relative intensity of scattered light (increasing toward the right on the abscissa of Figs. 9.1-9.3). Furthermore, during the 1,000-sec collection period, the instrument accumulated microbubble counts over successive 4-sec intervals and, in each case, determined the separate scattered-light intensity levels above which no more than 50 and 5 microbubbles reached during the interval (see Fig. 9.2(B)). The averaged results for the 1,000-sec collection period are plotted in both three and two dimensions in Fig. 9.2 (with the total counts given by the 6-digit number).

It can be seen that the relative proportions of microbubbles falling below the "50-microbubble" and "5-microbubble" scattered-light intensity levels (Fig. 9.2) changed little over a period of more than 15 min (i.e., 1,000 sec); this finding indicates that little, if any, loss of microbubbles due to (rapid) flotation occurred during this time period. Close inspection of part B of Fig. 9.2, however, does reveal a slight skewing of the microbubble population toward smaller sizes during the 1,000-sec period.

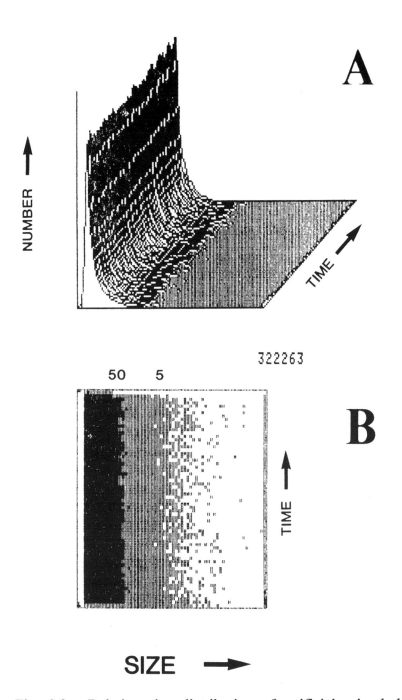

Fig. 9.2. Relative size distribution of artificial microbubbles in an unstirred medium. Light-scatter signals were collected over a total period of 1,000 sec, and plotted in three (part A) and two (part B) dimensions. (See text for further discussion.)

158

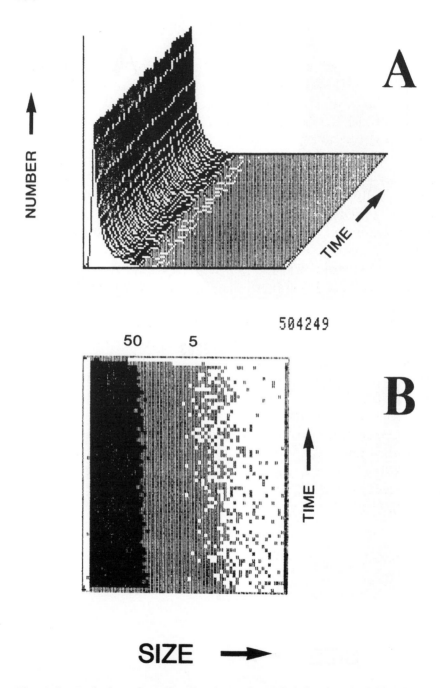

504249

Fig. 9.3. Relative size distribution of artificial microbubbles in a stirred medium. Light-scatter signals were collected over a total period of 1,500 sec, and plotted in three (part A) and two (part B) dimensions. (See text for further discussion.)

9.5   MICROBUBBLE PERSISTENCE WITH TIME

An effort was made to evaluate the possibility that the slight skewing of the artificial microbubble population (in Fig. 9.2) with time might be due to a continuing, even though very slow, dissolution of the newly formed, surfactant-coated microbubbles. Unlike the experiment referred to in Fig. 9.2, gentle automatic swirling of the preshaken (Filmix 3) surfactant stock solution was now performed as it was fed into the flow cell (see Section 9.2) of the Coulter EPICS V System.   Data collection and analysis were the same as in Section 9.4, except that the total collection period was now extended to 1,500 sec and each collection interval was 6 sec long.   The averaged results for the 1,500-sec collection period used in this experiment are plotted in both three and two dimensions in Fig. 9.3 (with the total counts given by the 6-digit number).

It can be seen that with the constant, gentle swirling of the surfactant solution in this experiment, the relative size of microbubbles falling below either the "50-microbubble" or the "5-microbubble" scattered-light intensity levels (Fig. 9.3) decreased more noticeably during the entire collection period (i.e., 25 min). This finding indicates that very slow dissolution of the newly formed, surfactant-coated microbubbles does continue, at least during the first half hour of their existence. Moreover, this gradual rate of dissolution of the artificial microbubbles can apparently be increased somewhat by circulation of the liquid.

Chapter 10

CONCENTRATED GAS-IN-LIQUID EMULSIONS IN ARTI-
FICIAL MEDIA. II. CHARACTERIZATION BY PHOTON
CORRELATION SPECTROSCOPY

10.1 BROWNIAN MOTION AND AUTOCORRELATION
ANALYSIS OF SCATTERED LIGHT INTENSITY

This section contains a general description of the principles by which the Coulter Model N4 Sub-Micron Particle Analyzer, used in this study to characterize artificial gas-in-water emulsions (see Section 10.4), determines sample particle size. The measuring principles are based on the theory of Brownian motion and photon correlation spectroscopy (ref. 464,465; see also Sections 10.2 and 10.4).

The motion caused by thermal agitation and the random striking of particles in a liquid by the molecules of that liquid is called Brownian motion. This molecular striking results in a vibratory movement that causes suspended particles to diffuse throughout a liquid. If the colloidal particles can be assumed to be approximately spherical, then for a liquid of given viscosity (η), at a constant temperature (T), the rate of diffusion, or diffusion coefficient (D) is inversely related to the particle size according to the Stokes-Einstein relation (ref. 126):

$$D = kT/(3\pi\eta d) \tag{1}$$

where k is Boltzmann's constant and d is the diameter of the particle. Accordingly, information concerning the size of particles in the liquid can be obtained by measuring the diffusion coefficients of the particles in the liquid.

The constantly changing patterns of suspended particles while in Brownian motion can be analyzed by light-scattering spectroscopy. The Coulter Model N4 measures time-dependent fluctuations in scattered laser light from a sample suspension to determine the sample particle diffusion coefficient. The scattering angle for the single-angle instrument used in this study was 90°. A photomultiplier (photon counter) detects fluctuations in the intensity of light scatter as a function of time while the particles undergo Brownian motion. These intensity fluctuations occur on the order of milliseconds, or less, depending on particle size. Photon correlation spectroscopy provides the means to analyze these intensity fluctuations which are characteristic of particles suspended in a liquid.

A convenient and accurate analysis of these intensity fluctuations is performed by computing the auto-correlation function of the measured intensities, $G(\tau)$, where

$$G(\tau) = \overline{I(t) \times I(t + \tau)} . \tag{2}$$

$G(\tau)$ is a time average of the scatter intensity at time t, I(t), times the intensity at a time τ later, or $I(t + \tau)$ where τ is on the order of msec or μsec. The autocorrelation function is determined by the Coulter Model N4 for a number of values of τ simultaneously, and the data is plotted as $G(\tau)$ versus τ. If the particles in the liquid are the same size and shape (i.e., spherical), and hence monodisperse, the curve of the autocorrelation function of the scattered light intensity is a single decaying exponential:

$$G(\tau) = \exp(-DK^2\tau) . \tag{3}$$

The decay rate of the exponential is $DK^2\tau$, where D is the translational diffusion coefficient and

$$K = (4\pi n / \lambda) \sin(\theta/2) . \tag{4}$$

K represents the following constant parameters: n is the index of refraction of the liquid, λ is the laser wavelength in air, and θ is the angle at which the scattering intensity is measured. For polydisperse samples, the autocorrelation function plot is the sum of exponentials for each size range. Once the average translational diffusion coefficient of the sample is determined, the equivalent spherical diameter can be determined by using the Stokes-Einstein relation (see ref. 466 and Section 10.2 for further discussion).

The analysis of the autocorrelation function data by the Coulter Model N4 is carried out by the Size Distribution Program (SDP), which gives the particle size distribution in the form of various output displays (see Section 10.4). The SDP analysis utilizes the computer program CONTIN developed by S.W. Provencher (ref. 467-470; see also Section 10.2). (This program has been tested on computer-generated data, monomodal polystyrene samples, and a vesicle system (ref. 466-468,471).) Since the SDP does not fit to any specific distribution type, it offers the ability to detect multimodal and very broad distributions.

10.2 BACKGROUND OBSERVATIONS ON MICELLAR GROWTH

Because of the above-mentioned multimodal detection capability of the CONTIN computer program, photon correlation spectroscopy has been used effectively by Flamberg and Pecora (ref. 466) to detect translational and rotational motions of micelles in solution. The results obtained can best be explained by a micellar growth model.

As mentioned by these authors, it had been argued earlier by numerous investigators that under certain conditions (for example, with many ionic surfactants in high concentrations at high ionic strengths and low temperatures) micelles tend to grow (ref. 450,472-478). In most of these cases, the micelle growth is from assumed small spherical micelles to large rod-shaped micelles (although some surfactant systems appear to favor an oblate shape) (ref. 450,466). These conclusions have been drawn from many physical measurements including diffusion studies, viscosity, NMR, and light scattering (ref. 466,479).

Additional evidence for the formation of such large, rodlike micelles, obtained using photon correlation spectroscopy and specifically the CONTIN computer program, is detailed in the paper by Flamberg and Pecora (ref. 466). As explained by these authors, the temporal fluctuations in the light scattering intensity obtained from large rod-shaped particles, as opposed to small or spherical particles, are more complex and depend not only on the particle's translational diffusion but also on its rotational diffusion. Photon correlation spectroscopy is capable of measuring rotational diffusion and, at the same time, offers several advantages. (Other methods for measuring rotational diffusion, such as dielectric relaxation (ref. 480), transient electric birefringence (ref. 481), and fluorescence depolarization are perturbing to the system. The dynamic light scattering technique of photon correlation spectroscopy does not require the application of strong external fields or the addition of any foreign material as a probe which may distort the micelles. The visible light used is not significantly absorbed and does not induce chemical reactions for most micellar systems studied (ref. 466).) For nonspherical particles with a long dimension ≥ 150 nm, the scattered light intensity time-autocorrelation function (see Section 10.1) becomes a sum of two (or more) discrete exponential decays. The first (slower) decay time is related to the translational diffusion coefficient term (from which one calculates, for a nonspherical particle, what is often referred to as the "apparent" hydrodynamic radius). If one assumes the particles are rigid rods, the second (faster) decay time represents a combination of traslational and rotational diffusion effects (ref. 466). (For nonrigid rods, the second (faster) decay time is essentially related to the rotational diffusion and/or to internal flexing of the particles (cf. ref. 466). Many investigators believe that rodlike micelles are actually flexible (ref. 473,476,479,482), although some do not (ref. 483).)

Flamberg and Pecora further point out that if the particle system consisted of monodisperse rods, then the first-order correlation function of the scattered light intensity could be fitted to a sum of two exponential decay terms. Unfortunately, micelles seem to be polydisperse. Instead of one or two discrete exponential decays, the light scattering data consist of a distribution of decays. To transform the autocorrelation function into distributions of particle size or diffusion coefficients, the CONTIN computer program was used (ref. 466). (This computational method was chosen by Flamberg and Pecora because it appears to give more consistent solutions even with the presence of limited amounts of dust (which gives a long-time exponential decay) and is easier to use than other methods such as the histogram method (ref. 484). Flamberg and Pecora also comment that the traditional method of analyzing polydispersity in dynamic light scattering, the method of cumulants (ref. 485), is successful only for monomodal distributions with no background dust scattering, and could not be used for their measurements (ref. 466).)

Experiments were carried out by Flamberg and Pecora on aqueous solutions of the ionic surfactant dodecyldimethylammonium chloride, in high concentration, and which contained 4.0 M NaCl at a temperature of 25°C. The light scattering data collected and then analyzed by the CONTIN program were plotted as particle distributions in terms of apparent hydrodynamic radii. (As mentioned in Section 10.1, the CONTIN program is given no a priori information as to the number of modes in the distribution. The program user only specifies the size range over which the solution will be given, but the peak locations seem to be insensitive to this. Control experiments with polystyrene spheres gave essentially the same results using a range of either 20-200 nm or 20-300 nm (ref. 466).) Using a scattering angle of 30° with the above-mentioned micellar system, a single peak was observed by Flamberg and Pecora which was related to the translational diffusion coefficient term (see above and Section 10.1). As the scattering angle was increased to 70°, a second peak appeared at a faster decay time (corresponding to a smaller apparent hydrodynamic radius). When the scattering angle was increased even more (to 90° and 110°), the relative amplitude of the faster peak increased. Flamberg and Pecora concluded that these observations indicated the presence of large, rodlike micelles.

Other ionic surfactants displaying sphere-rod equilibria of micelles dependent on micelle concentration include sodium dodecyl sulfate and various other dodecyldimethylammonium halides in concentrated NaCl and NaBr solutions (ref. 472,473,475,477, 478). Small spherical micelles are formed at the critical micelle concentration, and with increasing micelle concentration they associate together into large rodlike micelles (ref. 479). Such a sphere-rod equilibrium of ionic micelles occurs when the ionic strength exceeds a threshold value characteristic for the surfactant species. The threshold ionic strength is usually quite high. This indicates that the electrostatic effects of ionic micelles are sufficiently suppressed by the added salt so that the micelles behave as if they are essentially uncharged (ref. 479).

Accordingly, it is both understandable and to be expected that while most nonionic surfactants form spherical micelles alone in aqueous solutions above the critical micelle concentration, some nonionic surfactants can further associate into large or rodlike micelles, either at room or elevated temperatures, when the surfactant concentration exceeds the critical micelle concentration (ref. 486-493). For example, Imae and Ikeda (ref. 479) point out that hexaoxyethylene dodecyl ether (ref. 486,487,492) and hepta-oxyethylene cetyl ether (ref. 488,489,493) are found to form large micelles in water at room temperature. Nonionic surfactants can form rodlike micelles if the hydrophobicity is much stronger than the hydrophilicity. Imae and Ikeda believe that these properties of a nonionic surfactant can be determined by the length of the hydrocarbon chain relative to the size and hydration of the polar head group (ref. 479).

To provide a specific example, Imae and Ikeda (ref. 479) state that amine oxide is very hydrophilic and can constitute a good polar head group for nonionic surfactants at neutral pH. Dimethyl-dodecylamine oxide was first prepared by Hoh et al. (ref. 494), and its surface-active properties in aqueous solutions were investigated by measurements of surface tension (ref. 495), light scattering (ref. 496-498), and hydrodynamic properties (ref. 499). It was found that dimethyldodecylamine oxide can form only spherical micelles in water and aqueous NaCl solutions, when the micelle concentration is dilute (ref. 496,498). Similarly, the homolog dimethyltetradecylamine oxide forms only spherical micelles in water (ref. 496).

However, Imae and Ikeda (ref. 479) report the formation of rodlike micelles of the surfactant dimethyloleylamine oxide, $CH_3(CH_2)_7CH=CH(CH_2)_8N(CH_3)_2O$, in aqueous NaCl solutions. Without added acid, the surfactant behaves as a nonionic (ref. 479). These authors measured the angular dependence of light scattering from solutions of the surfactant in the presence of NaCl from 5 x 10^{-4} M to 0.1 M at 25°C. The calculated molecular weight and radius of gyration of micelles increase with increasing micelle concentration and reach constant values, indicating occurrence of a sphere-rod equilibrium dependent on the micelle concentration. With increasing NaCl concentration, rod-shaped micelles are larger in molecular weight and become longer. The micelles formed at NaCl concentrations higher than 10^{-3} M are nearly monodisperse rods when the micelle concentration is high. Specifically, the rod-like micelles of dimethyloleylamine oxide in 10^{-2} M and 5 x 10^{-2} M NaCl solutions have molecular weights of 4,760,000 and 6,900,000, respectively, and behave as semiflexible or wormlike chains. In 5 x 10^{-2} M NaCl, they have an end-to-end distance of 380 nm. The large aggregation number of the rodlike micelles is induced by the strong cohesion of long hydrocarbon chains (ref. 479).

In summary, while the nonionic surfactant dimethyldodecylamine oxide forms only spherical micelles even in 0.20 M NaCl (see above and ref. 498), micelles of dimethyloleylamine oxide are subject to a sphere-rod equilibrium in aqueous solutions of NaCl as dilute as 10^{-4} M and even in water alone (ref. 500). Thus, Imae and Ikeda (ref. 479) conclude the rodlike micelles are stabilized more, as compared with the spherical micelles, when the hydrocarbon chain of the surfactant molecule is longer. This conclusion is, therefore, consistent with the earlier-mentioned belief of these authors that the rodlike micelles are more stable when the polar head group of the surfactant molecule is smaller and the chain length of its hydrocarbon part is longer (ref. 473). Since the surfactants referred to all behave as nonionics, these findings of rodlike micelle production have direct relevance to the formation of artificial gas microbubbles (see Section 10.3) with either the earlier-mentioned surfactant mixture Filmix 3 (see Chapter 9 and Section 10.4) or another, related surfactant preparation (see Section 10.4).

10.3   SOLUBILIZATION OF GASES IN MICELLES

In the preceding section, the importance of long hydrocarbon chains of surfactants to the stability of large rodlike micelles was reviewed. At the same time, however, these hydrocarbon chains can also serve effectively for the solubilization of various gases within such micelles.

It is well known that surfactants dissolved in aqueous solutions serve to enhance the solubility of ordinarily insoluble organic compounds, both solids and liquids. This phenomenon, commonly referred to as solubilization (ref. 501), has important commercial applications and as a consequence, has been the subject of considerable research (for reviews see ref. 450,502-504). Yet, as King and co-workers point out (ref. 501,505,506), less well known is the long-recognized fact that micellar solutions of surfactants are also capable of solubilizing gases (and vapors of low-molecular-weight compounds) in much the same manner as solids and liquids (ref. 501,505-518); this aspect of micellar solubilization has not received as much attention over the years.

Recently, a series of experiments have been performed by King and co-workers in which micellar solubilities of a variety of gases were determined under a range of experimental conditions (ref. 501,505,506,514,515,517). The gases used had widely differing solubilities, i.e., He, O_2 , Ar, CH_4 , C_2H_6, and C_3H_8 , and were tested at 25°C in solutions of either anionic or cationic surfactants (ref. 501,505). It was found that the solubility of each gas followed Henry's law at all surfactant concentrations. Further tests with several of these gases revealed that below the critical micelle concentration, the solubility of a given gas was nearly independent of surfactant concentration and approximated the corresponding solubility in water. Above the critical micelle concentration, the solubility of each gas increased linearly with surfactant concentration indicating that micelles act to solubilize gas molecules (ref. 501).

Since the series of gases used are all nonpolar, these authors argue that one should expect an increasing contribution of London dispersion forces to the overall solute-solvent potential energy of interaction as one proceeds across the series in the direction: O_2 , CH_4 , C_2H_6 , C_3H_8 (i.e., those gases employed in the later experiments (ref. 501)). Thus, for systems involving nonpolar solvents for which nondirectional dispersion forces constitute the dominant mode of interaction, one expects gas solubility to increase going from O_2 to C_3H_8 . This was observed to be the case for the gases dissolved in micelles and hydrocarbons, but not in the case of water for which entropic effects due to hydrogen bonding exert a large effect on solubility. Bolden et al. (ref. 501) conclude that this finding suggests that the microenvironment surrounding gas molecules solubilized within micelles, composed of the various surfactants used in their study, more closely resembles a nonpolar hydrocarbon than water. Furthermore, their data clearly demonstrated that each gas was much more soluble in micelles than in water, as would be expected if the site of solubilization within a micelle were hydrocarbon-like (ref. 501).

Accordingly, this probable site of gas solubilization suggests that the much higher solubility of nonpolar gases in micelles than in water ought to be observed in both spherical and rod-shaped micelles, composed of either ionic or nonionic surfactants. Furthermore, the very large rodlike micelles (up to at least 380 nm in length) formed by nonionic surfactants, as reviewed in Section 10.2, would each be expected to solubilize a relatively large quantity of gas compared to an ordinary, small spherical micelle. The same large gas-solubilization capacity is to be expected of any large, rodlike micelles formed from the nonionic surfactants comprising the earlier-described surfactant mixture used to produce artificial gas microbubbles (see Chapter 9). (Although the CAV-CON Filmix 3 surfactant mixture referred to here does contain cholesterol and (nonionic) cholesterol derivatives besides saturated monoglycerides, gas solubilization within the mixed micelles formed should still be very appreciable; specifically, it has been reported by Miller et al. that introduction of about 30 mole % cholesterol into acyl lipid bilayers caused only a slight reduction in the solubilities of 8 nonpolar gases in the bilayers (ref. 513).) Finally, the gas-rich interior of these possible large, rodlike micelles of artificial microbubble surfactant might well reach a state of supersaturation if, during the flexing motions characteristic of large rodlike micelles (ref. 466,479), appreciable water penetration into the micelle were to occur (cf. ref. 519,520). In such a situation, the combined factors of gas supersaturation (and hence impending phase separation), a low surface tension, and an enveloping surfactant monolayer (which in turn is surrounded by the saturated, aqueous surfactant solution) together could reasonably be expected to lead to microbubble formation. In the next section, particle size distributions, derived from photon correlation spectroscopy of microbubble surfactant solutions, are presented which do indeed suggest the formation of gas microbubbles from large micellar structures, as well as the reverse process of collapse of gas microbubbles into such micellar structures.

10.4 SIZE DISTRIBUTION OF SYNTHETIC MICRO-BUBBLES: FORMATION, COALESCENCE, FISSION, AND DISAPPEARANCE

10.4.1 Bimodal size distribution of the microbubble-surfactant particle population

The Coulter Model N4 Sub-Micron Particle Analyzer, used to obtain the measurements in this section on artificial gas-in-water emulsions, was described earlier (see Section 10.1). Much mention has already been made of various advantageous features of the CONTIN computer program used for the particle size analysis performed by the Coulter Model N4 (see Section 10.1), and employed in the earlier-mentioned photon correlation spectroscopy work of Flamberg and Pecora (see Section 10.2).

Another noteworthy feature, as stressed by Flamberg and Pecora, is that the smoothing algorithm in the CONTIN program has a tendency to give apparent distributions that are broader than the actual distributions. This slight oversmoothing reduces the possibility of spurious modes being reported. Hence, it is unlikely that the CONTIN program would report two modes when only one is present. When the signal-to-noise ratio in the data is high, the locations of the two peaks (in terms of their apparent hydrodynamic sizes) given by CONTIN are reliable (ref. 466). This point is significant because most of the particle-size data presented in this section are bimodal in nature (see below). (However, as will be discussed below, this bimodal feature of the data most probably does not result from effects of rotational motion of the particles detected by the Coulter Model N4; briefly, besides the rather large values of the equivalent spherical diameters and/or apparent hydrodynamic diameters computed, there is the added reason that the software of the Coulter Model N4 is configured to favor blanking out of secondary signals derived from rotational motion.)

Measurements were first taken on samples of a saturated, aqueous solution of the same artificial microbubble surfactant mixture (Filmix 3) used for the gas-in-water emulsion experiments described in Chapter 9. As before, the saturated aqueous solution of the surfactant mixture (prefiltered through a 1.2-μm pore-diameter membrane filter) was simply shaken vigorously for a few seconds, in an air atmosphere at room temperature, to form the concentrated gas-in-water emulsions. Fig. 10.1 shows the particle size distributions obtained by means of photon correlation spectroscopy using the CONTIN computer program. The first distribution, labeled sample 7, was obtained by collecting scatter emissions for a total period of 480 sec (at 20°C) beginning 5.0 min after shaking ended. It can be seen from this record that a bimodal distribution of particle sizes was obtained. It seemed plausible that the smaller and more numerous particles detected, possessing an average (apparent) hydrodynamic diameter of 139 nm, were large

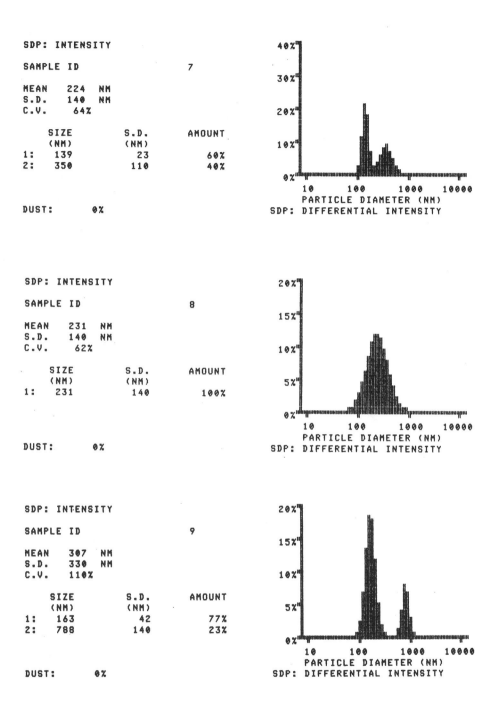

Fig. 10.1. Particle size distribution determined for artificial, surfactant-stabilized microbubbles (and micelles) in distilled water.

rodlike micelles (see Section 10.2) of the nonionic surfactants used. Hence, one might then logically hypothesize that the larger and less numerous particles detected, possessing an average hydrodynamic diameter of 350 nm, represent newly formed, surfactant-stabilized gas microbubbles. To begin testing this hypothesis, the solution was allowed to remain undisturbed overnight in the cuvette and another measurement (again 480 sec in duration at 20°C) taken the following day. The particle size distribution obtained, i.e., sample 8 in Fig. 10.1, was now monomodal, with an average hydrodynamic diameter of 231 nm. This value was within the range frequently reported for rodlike micelles (see Section 10.2). The absence of a second peak with larger particles (probably representing microbubbles) seemed consistent, at first, with the tentative explanation that such microbubbles were lost by flotation during the prolonged quiescent period. Upon reshaking the solution and beginning the 480-sec data collection period immediately thereafter, the particle size distribution obtained (sample 9 in Fig. 10.1) was indeed again bimodal. While the smaller particles detected (again probably rodlike micelles) possessed an average size within a standard deviation of the presumed (rodlike) micellar particles measured in sample 7 earlier, the average size of the larger particles (probably microbubbles) differed by more than a factor of two.

To evaluate the interrelationship between the suspected micellar and microbubble populations in more detail, a fresh, saturated solution of the same artificial microbubble surfactant mixture was prepared (and filtered through a 6.0-μm pore-diameter membrane filter). For this series of measurements, the surfactant solution was shaken (1 oscillation/sec for 5 sec) only at the beginning of the experiment and the 300-sec data collection periods (at 20°C) were begun at progressively longer time intervals thereafter. Specifically, in Figs. 10.2 and 10.3, data collection for samples 23-26 began at 2, 10, 21, and 30 min, respectively, after shaking ended. It can be seen that the particle size distributions obtained were bimodal for all four samples (i.e., samples 23-26). Interestingly, the sequence of records representing samples 23-25 (Fig. 10.2) displays a progressive and simultaneous increase in the average hydrodynamic diameter of both the suspected micellar and microbubble populations. A possible explanation for this trend is that as increasingly large, rodlike, gas-rich micelles are formed, an

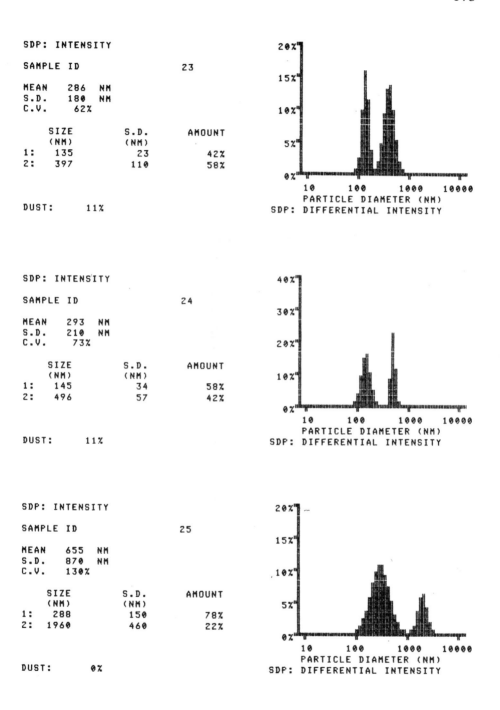

Fig. 10.2. Particle size distribution determined for artificial, surfactant-stabilized microbubbles (and micelles) in distilled water.

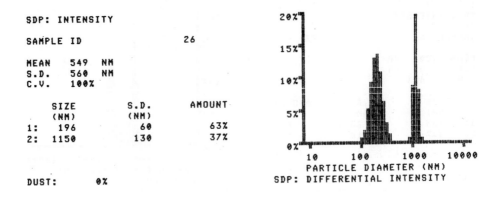

```
SDP: INTENSITY                              20%

SAMPLE ID                26             15%

MEAN    549  NM
S.D.    560  NM
C.V.    100%                            10%

        SIZE        S.D.      AMOUNT
        (NM)        (NM)                5%
    1:   196         60        63%
    2:  1150        130        37%      0%
                                           10      100     1000    10000
                                           PARTICLE DIAMETER (NM)
    DUST:        0%                     SDP: DIFFERENTIAL INTENSITY
```

Fig. 10.3. Particle size distribution determined for artificial, surfactant-stabilized microbubbles (and micelles) in distilled water.

increased generation of surfactant-stabilized gas microbubbles can take place (see Section 10.3); these newly formed microbubbles may then coalesce to form several-fold larger diameter micro-bubbles. Indirect support for hypothesizing such coalescence of microbubbles is received from the fact that, in samples 23-25 (Fig. 10.2), the proportion of suspected microbubbles is seen to progressively decrease (i.e., from 58% to 22%) relative to suspected micellar particles as the microbubble diameter increases (i.e., from 397 nm to 1,960 nm). In addition, the particle size distribution obtained for the subsequent sample 26 (Fig. 10.3) suggests that the trend towards coalescence may be reversible, since now the average hydrodynamic diameters of the micellar and microbubble populations simultaneously decrease. In this case, mi-crobubble fission may now have occurred with the surfactant molecules needed for the expanding (total) gas/water interface being obtained from the rodlike micelles. This explanation is consistent with the fact that the proportion of micelles is seen to decrease relative to the microbubble population as microbubble diameter decreases. (It is also consistent with the independent, earlier experimental observation that gas microbubble fission readily occurs in some physiological and other solutions where high concentrations of surfactants are present (ref. 521), i.e., well above the critical micelle concentration. In addition, it may be relevant to include a more general observation that this entire dynamic interrelationship, between the micellar and microbubble populations, bears many similarities to reports that between the regions of micellar solution and liquid/liquid microemulsion in some multicomponent phase diagrams, a zone of instability has been revealed through analysis of particle sizes using photon correlation spectroscopy (ref. 522, 523).)

10.4.2 Combined evidence that the larger-diameter Filmix particles (i.e., subpopulation) are surfactant-stabilized gas microbubbles

Several arguments in favor of the belief that the population of larger-diameter particles, detected in these photon correlation spectroscopy measurements on artificial microbubble surfactant solutions, are indeed microbubbles (cf. Chapter 12) are as follows:

1) The surfactant mixture used (CAV-CON Filmix 3) is identical to that used to form the artificial gas-in-water emulsions described in Chapter 9, where the concentrated emulsion particles observed were all above 0.3 µm in diameter (which is the lower detection limit of the laser-based flow cytometer instrument) and therefore did not include the co-existing micelles.

2) The surfactant solution was passed through a precleaned, membrane filter which removed all solid debris greater than 6.0 µm in diameter.

3) Since none of the surfactants used are liquids, oil-in-water microemulsions or emulsions cannot be the basis for the existence of the larger-diameter population (or even the smaller-diameter population) of particles detected.

4) In accompanying decompression tests, it was found, in the course of developing Filmix 3, that those surfactant solutions which produced the higher concentrations of growing bubbles upon decompression (to below 1 atm) similarly produced the greater degree of light scatter in the absence of decompression.

5) The well-documented ability of Filmix particles to concentrate rapidly and selectively in tumor tissue (ref. 524-535), with no accumulation in surrounding normal tissue, has been found (ref. 529) to cause a significant change in the "bulk magnetic susceptibility" (BMS) inside tumors in vivo as detected by magnetic resonance imaging (MRI). This observed change in the BMS within the tumor tissue could only occur if the larger-diameter Filmix lipid particles did, in fact, represent "lipid-coated microbubbles" (LCM) which contained bulk (paramagnetic) gas,

e.g., air. In other words, the MRI (Filmix-)contrast mechanism observed was that of a bulk-magnetic-susceptibility increase from the local accumulation of the air-filled LCM inside the tumor mass, i.e., from a significant increase of the local volume fraction of the gas microbubbles and therefore also the local BMS value (ref. 529). (See Section 12.4 below for more explanatory details.)

6) The larger Filmix particles (LCM) have been shown to also serve as a suitable sonographic contrast agent (ref. 524-526,528, 536), due to their gaseous core, causing increased intensity of echoes that persists for 30 min in tumors in vivo (ref. 525,526,528). (Microbubbles in general work by vibrating in an ultrasound beam, rapidly contracting and expanding in response to the pressure changes of the sound wave. In theory, bubbles of the proper size resonate strongly with sound to provide the brightest image, i.e., higher echo intensities. To a first approximation, the physical relation between ultrasound transducer frequency and radius of the resonating bubble requires a diameter of less than 5 μm if the transducer frequency is in the medical imaging range of 1-10 MHz (ref. 444,445).) Using lipid-specific stains (ref. 526, 528), the distribution of microbubbles in the tumor and surrounding organ were compared to the distribution of echoes on the sonogram. It was found that the histological distribution of microbubbles correlated well with the spatial distribution of echo intensities, the measured signal-to-noise ratio of the echoes, and the spectral characteristics of the tumor echoes (ref. 526)(see also Section 12.2).

7) In a related ultrasound study it was found that when a higher diagnostic-ultrasound output intensity is used (e.g., continuous-wave ultrasound [0.25 W/cm^2]), the larger Filmix particles (LCM) vibrate more violently and often burst (ref. 530). This finding is consistent with the established fact (see ref. 537 for literature review) that gas pockets, e.g. microbubbles, are known to act as heat sinks as well as cavitation nuclei when exposed to continuous-wave ultrasound (ref. 538-543). Accordingly, in this particular animal study, experimental tumors were insonated after the LCM had accumulated within the tumor mass. From this treatment, the authors observed that: "Selective necrosis, lymphocyte proliferation and hemorrhage within the tumor can be demonstrated. Preliminary data are given to demonstrate this phenomenom. The mechanism of the effect is discussed in the context of both heating and cavitation" (ref. 530). (See Chapter 12 for further discussion.)

10.4.3 <u>Apparent reversible and/or cyclical behavior: Microbubble formation and coalescence versus microbubble fission and disappearance</u>

The next two series of photon correlation spectroscopy measurements were designed to examine the above-described reversible process, i.e., microbubble formation and coalescence followed by microbubble fission and disappearance, in more detail. A new batch of the Filmix 3 surfactant mixture was used to form a saturated aqueous solution, which was filtered through a 6.0 μm (pore diameter) membrane filter. As before, the surfactant solution was shaken (1 oscillation/sec for 5 sec) at the beginning of the experiment and the 300-sec data collection periods (at 20°C) commenced as follows: In Figs. 10.4-10.6, data collection for samples 27-35 began at 2, 11, 20, 30, 39, 50, 61, 70, and 80 min, respectively, after shaking ended. The surfactant solution was then reshaken and the experiment repeated in the following, similar manner: In Figs. 10.7-10.10, data collection periods of 180 sec each (at 20°) for samples 37-48 began at 2, 9, 17, 24, 31, 39, 47, 54, 62, 70, 77, and 85 min, respectively, after shaking ended. It can be seen from detailed inspection of Figs. 10.4-10.6 and Figs. 10.7-10.10 that the earlier-mentioned reversible process of microbubble formation and coalescence followed by microbubble fission and disappearance is actually part of a cyclical process. It appears that each cycle may end with the ultramicrobubbles, produced by microbubble fission, collapsing into the (presumably rodlike) micelles (cf. sample 47 in Fig. 10.10). (In addition, it is possible that occasionally either subpopulations of smaller-than-usual micelles are detected (see sample 30 in Fig. 10.5, sample 37 in Fig. 10.7, and sample 41 in Fig. 10.8) or the software of the Coulter Model N4 is not always able to blank out the effects of rotational motion at a scattering angle of 90°.) Also noteworthy is that sometimes (under the experimental conditions used for these particular tests, i.e., in a *distilled water* medium) a subpopulation of larger-than-usual microbubbles, resulting from prolonged coalescence, might occur (see sample 29 in Fig. 10.4); the detection of these large microbubbles (≥ 10 μm [in *distilled water*]) does not appear to be an artifact since the dust reading for this sample was 0% (see also below).)

As a control for all of the above experiments, measurements were made on a sample of distilled water, which had been passed

178

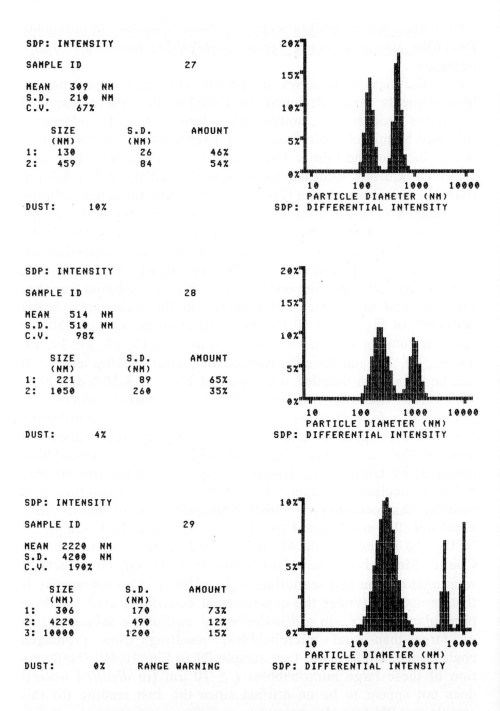

SDP: INTENSITY

SAMPLE ID   27

MEAN   309 NM
S.D.   210 NM
C.V.   67%

	SIZE (NM)	S.D. (NM)	AMOUNT
1:	130	26	46%
2:	459	84	54%

DUST: 10%

SDP: INTENSITY

SAMPLE ID   28

MEAN   514 NM
S.D.   510 NM
C.V.   98%

	SIZE (NM)	S.D. (NM)	AMOUNT
1:	221	89	65%
2:	1050	260	35%

DUST: 4%

SDP: INTENSITY

SAMPLE ID   29

MEAN   2220 NM
S.D.   4200 NM
C.V.   190%

	SIZE (NM)	S.D. (NM)	AMOUNT
1:	306	170	73%
2:	4220	490	12%
3:	10000	1200	15%

DUST: 0%   RANGE WARNING

Fig. 10.4. Particle size distribution determined for artificial, surfactant-stabilized microbubbles (and micelles) in distilled water.

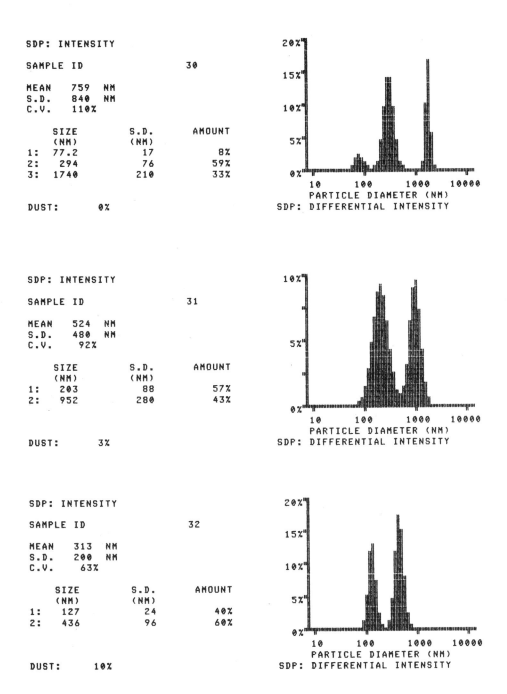

Fig. 10.5. Particle size distribution determined for artificial, surfactant-stabilized microbubbles (and micelles) in distilled water.

180

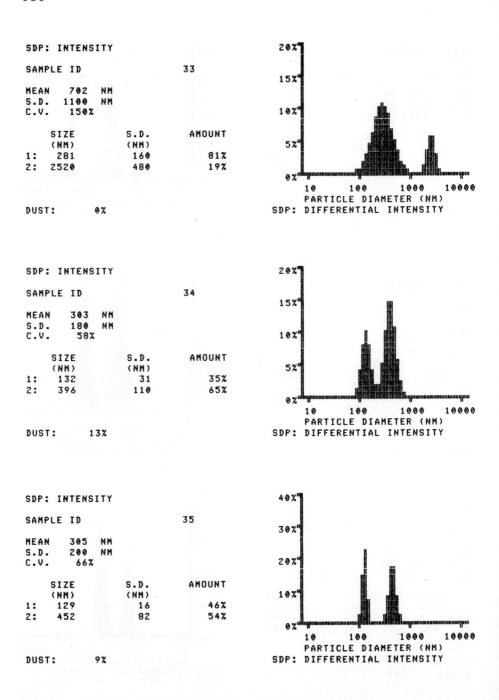

Fig. 10.6. Particle size distribution determined for artificial, surfactant-stabilized microbubbles (and micelles) in distilled water.

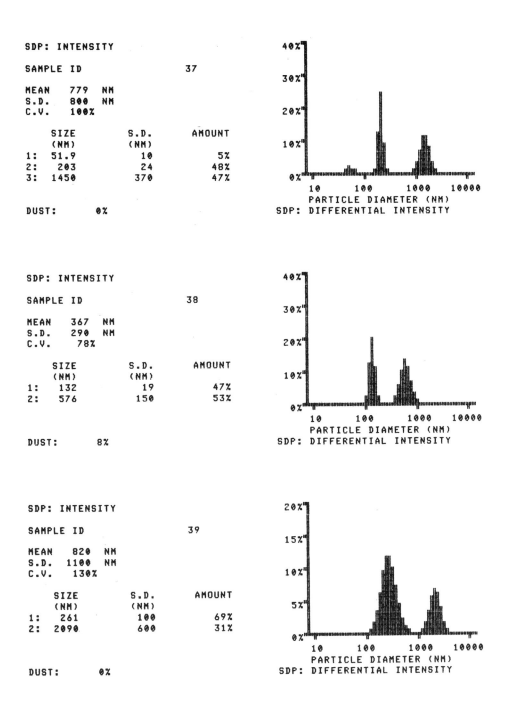

Fig. 10.7. Particle size distribution determined for artificial, surfactant-stabilized microbubbles (and micelles) in distilled water.

182

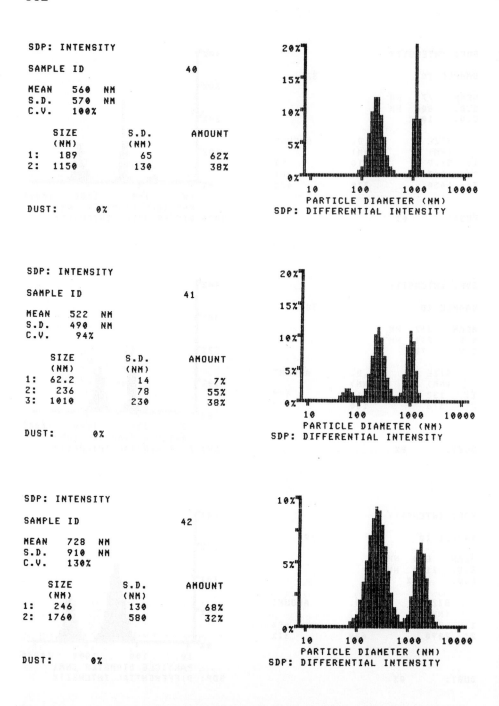

Fig. 10.8. Particle size distribution determined for artificial, surfactant-stabilized microbubbles (and micelles) in distilled water.

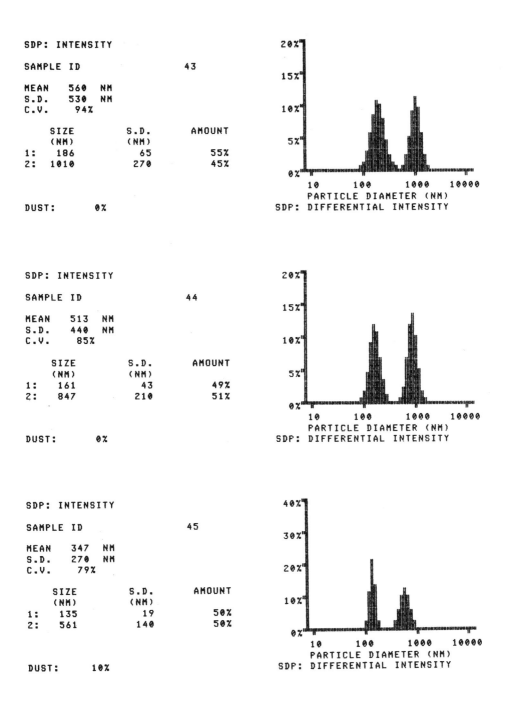

Fig. 10.9. Particle size distribution determined for artificial, surfactant-stabilized microbubbles (and micelles) in distilled water.

184

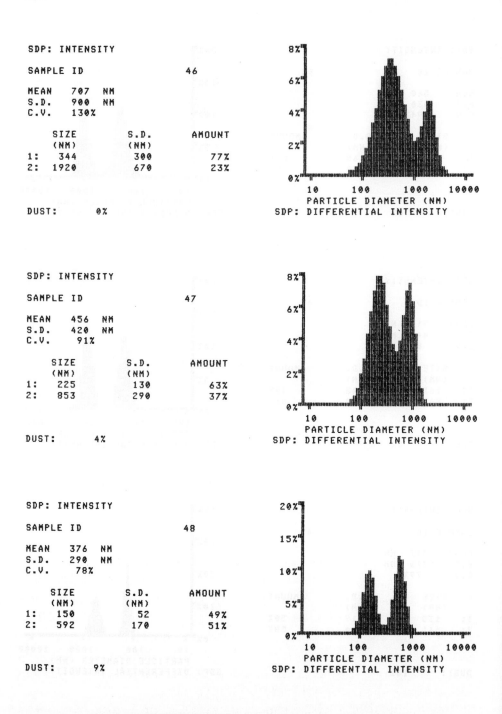

Fig. 10.10. Particle size distribution determined for artificial, surfactant-stabilized microbubbles (and micelles) in distilled water.

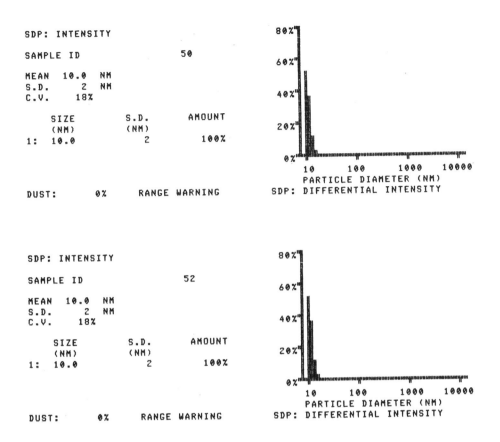

Fig. 10.11. "Apparent" particle size distribution determined for distilled water alone.

through a 6.0 μm filter and shaken as before. The results are shown in Fig. 10.11, where data collection periods of 180 sec each (at 20°C) for samples 50 and 52 commenced at 2 and 20 min, respectively, after shaking ended. It can be seen from Fig. 10.11 that the distribution of particle sizes was not bimodal. In both cases, all of the particles recorded are at the low end of the detection range, which is an artifact of the software arising from too low a concentration of detectable particles.

In the final series of measurements, a prolonged effort was made to record the cyclical process of microbubble formation/coalescence/fission/disappearance in considerable detail. The primary objective was to better determine if this cyclical micro-bubble process does actually complete each cycle with the ultramicrobubble population, produced by microbubble fission, collapsing into the micelle population. A secondary objective of this experiment was to confirm that the cyclical process occurred with at least one other artificial microbubble surfactant mixture. The surfactant mixture chosen, CAV-CON Filmix 2, had already been determined from decompression tests to be effective in the production of surfactant-stabilized microbubbles; this surfactant mixture contains approximately 30% (by weight) less mono-glycerides than does the earlier-mentioned mixture, i.e., Filmix 3. A saturated, aqueous solution of the Filmix 2 surfactant mixture was first filtered through a 6.0 μm filter and shaken as before. The results of the measurements are shown in Figs. 10.12-10.23, where the data collection periods of 180 sec each (at 20°) for samples 65-100 commenced at 2, 9, 17, 24, 31, 39, 47, 54, 62, 70, 77, 85, 92, 99, 107, 115, 123, 130, 138, 145, 153, 160, 167, 175, 183, 191, 198, 205, 213, 220, 228, 236, 246, 254, 261, and 269 min, respectively, after shaking ended. Inspection of Figs. 10.12-10.23 reveals that a saturated solution of Filmix 2 does, in fact, similarly display the cyclical process of microbubble formation/coales-cence/fission/disappearance. Moreover, enough sequential records were obtained to be able to find one instance of complete merging of the micelle and microbubble populations (see sample 81 in Fig. 10.17). (It was also noted that with this particular surfactant mixture [again in *distilled water*], prolonged coalescence of mi-crobubbles, resulting in subpopulations of larger-than-usual microbubbles, appeared to occur more frequently (see sample 68 in Fig. 10.13, sample 76 in Fig. 10.15, and sample 88 in Fig. 10.19).)

In conclusion, it is possible to prepare concentrated gas-in-water emulsions using various surfactant mixtures. The artificial, surfactant-stabilized microbubbles produced apparently undergo a cyclical process of microbubble formation/coalescence/fission/dis-appearance, where the end of each cycle is characterized by a collapse of the gas microbubbles into large micellar structures -- only to re-emerge soon after as newly formed, gas microbubbles. This cyclical microbubble process is promoted by prior mechanical agitation of, and hence entrapment of macroscopic gas bubbles in, these saturated surfactant solutions.

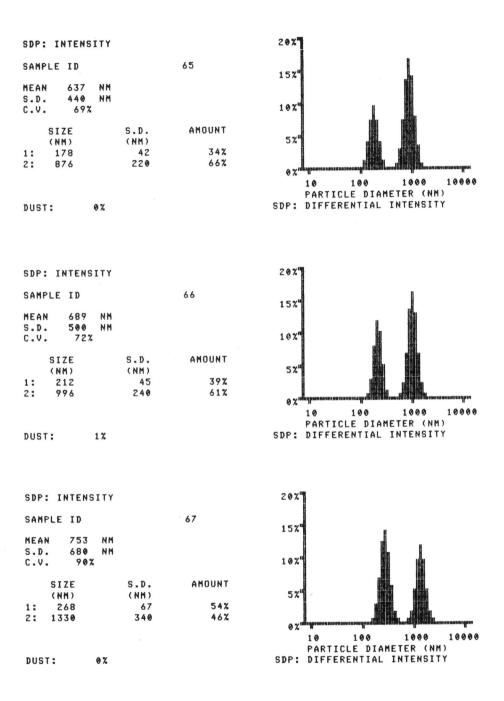

Fig. 10.12. Particle size distribution determined for artificial, surfactant-stabilized microbubbles (and micelles) in distilled water.

188

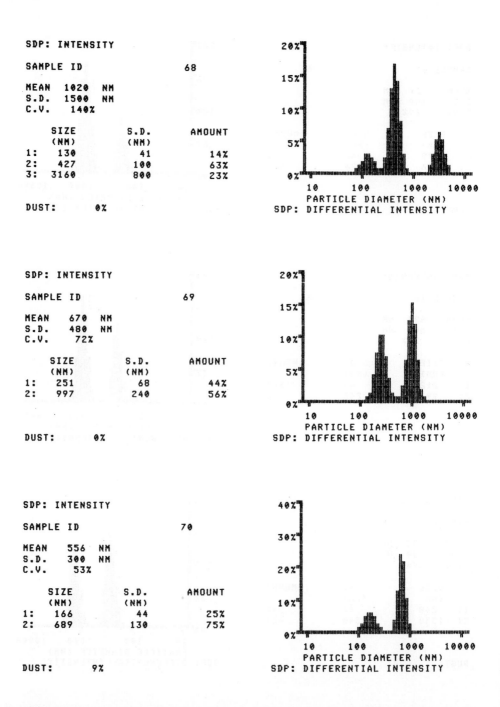

SDP: INTENSITY

SAMPLE ID   68

MEAN   1020 NM
S.D.   1500 NM
C.V.   140%

	SIZE (NM)	S.D. (NM)	AMOUNT
1:	130	41	14%
2:	427	100	63%
3:	3160	800	23%

DUST: 0%

SDP: DIFFERENTIAL INTENSITY

SDP: INTENSITY

SAMPLE ID   69

MEAN   670 NM
S.D.   480 NM
C.V.   72%

	SIZE (NM)	S.D. (NM)	AMOUNT
1:	251	68	44%
2:	997	240	56%

DUST: 0%

SDP: DIFFERENTIAL INTENSITY

SDP: INTENSITY

SAMPLE ID   70

MEAN   556 NM
S.D.   300 NM
C.V.   53%

	SIZE (NM)	S.D. (NM)	AMOUNT
1:	166	44	25%
2:	689	130	75%

DUST: 9%

SDP: DIFFERENTIAL INTENSITY

Fig. 10.13. Particle size distribution determined for artificial, surfactant-stabilized microbubbles (and micelles) in distilled water.

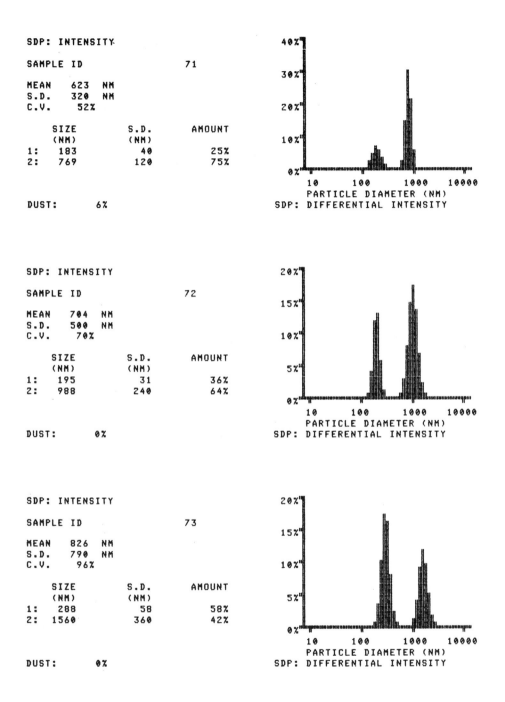

Fig. 10.14. Particle size distribution determined for artificial, surfactant-stabilized microbubbles (and micelles) in distilled water.

190

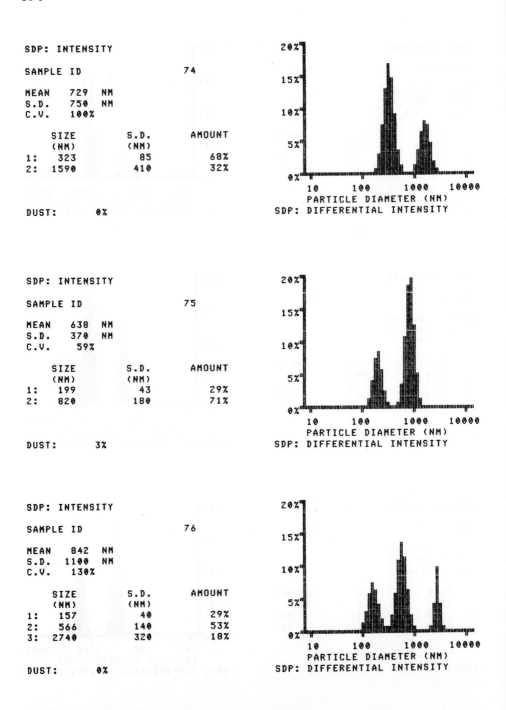

Fig. 10.15. Particle size distribution determined for artificial, surfactant-stabilized microbubbles (and micelles) in distilled water.

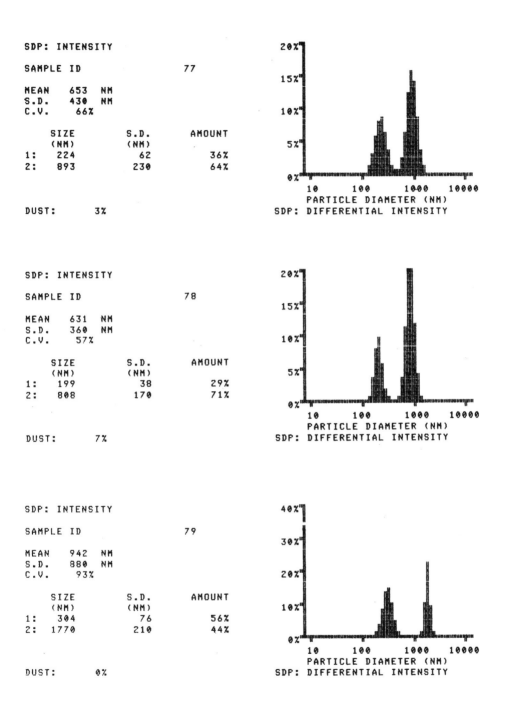

Fig. 10.16. Particle size distribution determined for artificial, surfactant-stabilized microbubbles (and micelles) in distilled water.

192

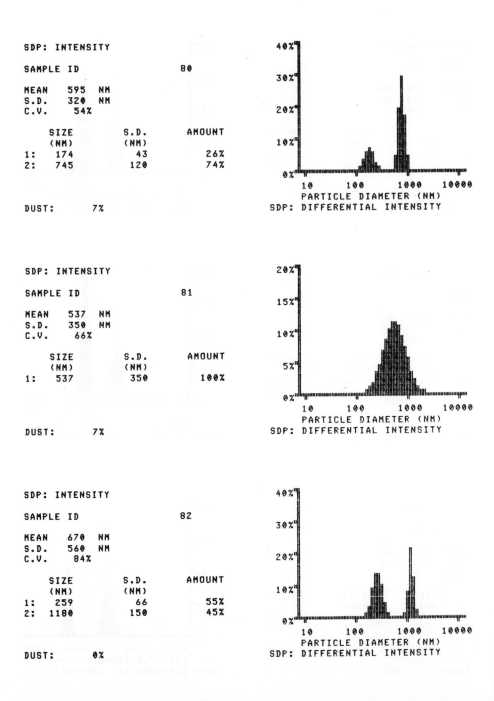

Fig. 10.17. Particle size distribution determined for artificial, surfactant-stabilized microbubbles (and micelles) in distilled water.

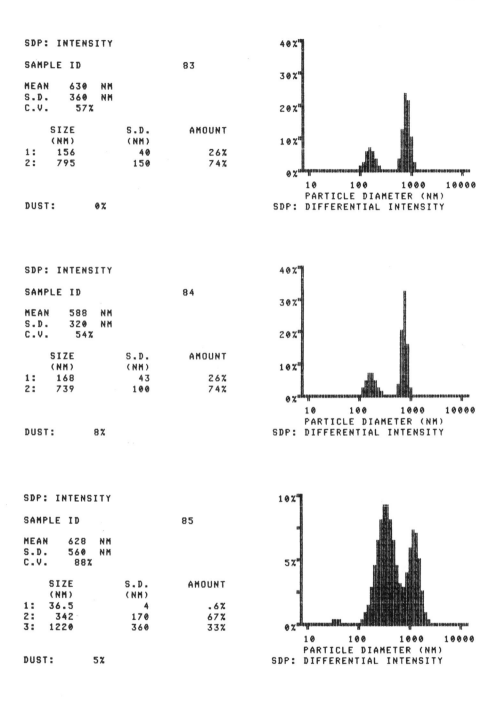

Fig. 10.18. Particle size distribution determined for artificial, surfactant-stabilized microbubbles (and micelles) in distilled water.

194

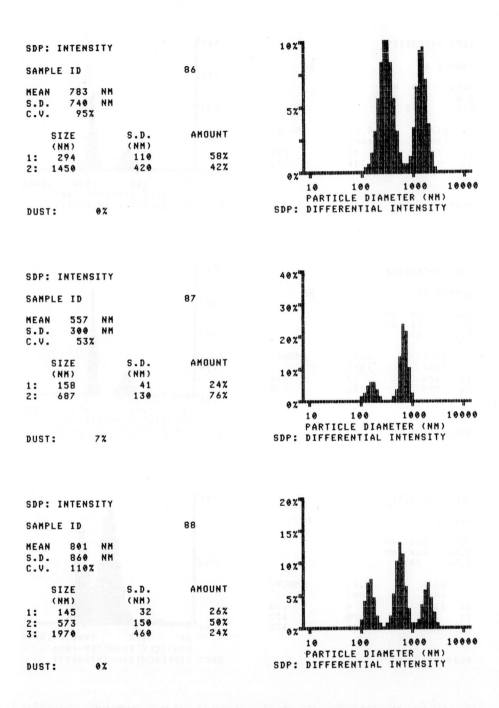

Fig. 10.19. Particle size distribution determined for artificial, surfactant-stabilized microbubbles (and micelles) in distilled water.

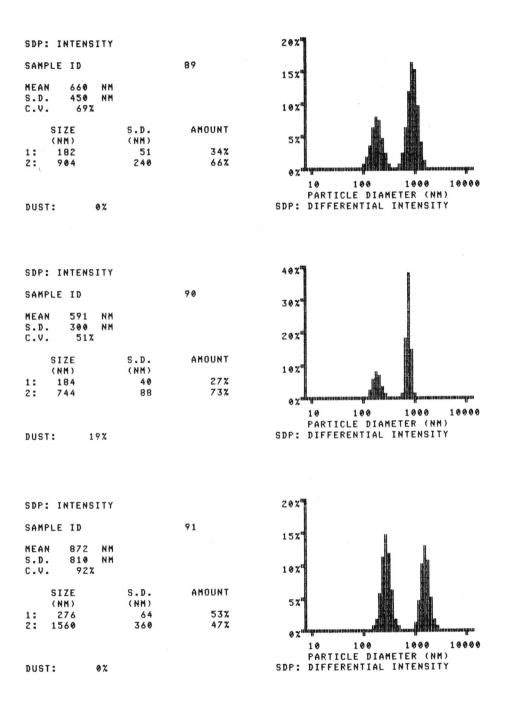

Fig. 10.20. Particle size distribution determined for artificial, surfactant-stabilized microbubbles (and micelles) in distilled water.

196

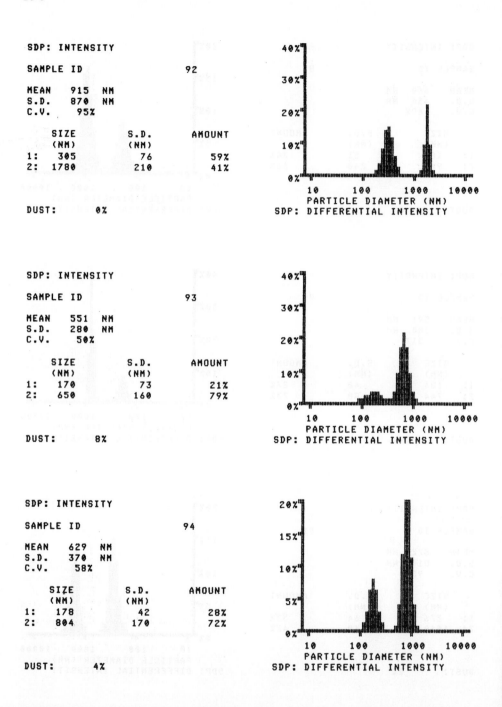

Fig. 10.21. Particle size distribution determined for artificial, surfactant-stabilized microbubbles (and micelles) in distilled water.

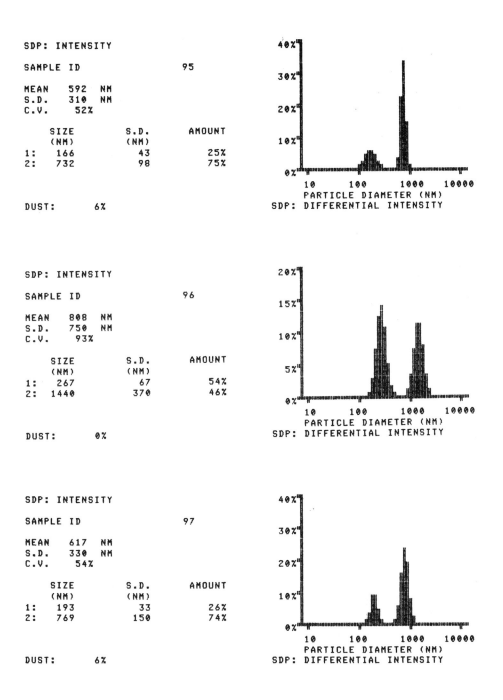

Fig. 10.22. Particle size distribution determined for artificial, surfactant-stabilized microbubbles (and micelles) in distilled water.

198

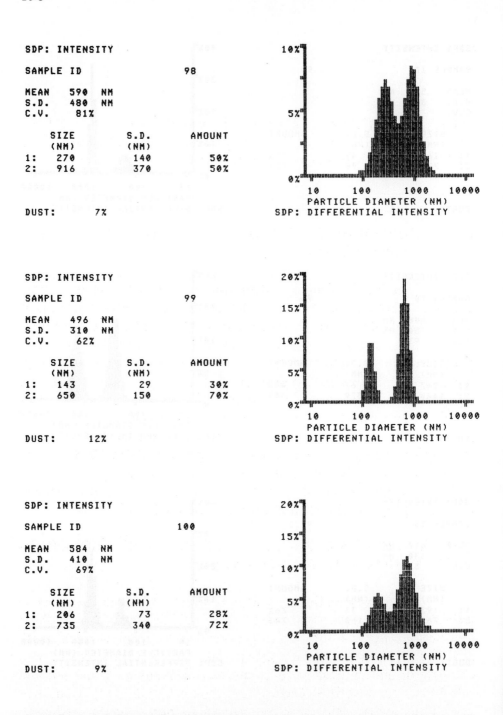

Fig. 10.23. Particle size distribution determined for artificial, surfactant-stabilized microbubbles (and micelles) in distilled water.

Chapter 11

CONCENTRATED GAS-IN-LIQUID EMULSIONS IN ARTIFI-
CIAL MEDIA. III. REVIEW OF MOLECULAR MECHANISMS
INVOLVED IN MICROBUBBLE STABILIZATION

11.1 MICROBUBBLE LONGEVITY AND INTERAGGRE-
GATE INTERACTIONS

From the data presented in Chapter 10, it becomes evident
that the extreme longevity of the artificial surfactant-stabilized
microbubbles described therein is, in part, related to their continuous interaction with the simultaneously formed "mixed micelle"
population in the saturated surfactant solution. More specifically,
the surfactant-stabilized microbubbles produced by mechanical
agitation of saturated solutions of either CAV-CON's Filmix 2 or
Filmix 3 apparently undergo a cyclical (or reversible) process of
microbubble formation/coalescence/fission/disappearance, where
the end of each cycle is characterized by a collapse of the lipid-
coated microbubbles into large micellar structures (i.e., rodlike
multimolecular aggregates), only to re-emerge soon after as newly
formed, lipid-coated microbubbles (see also below).

11.2 MOLECULAR PACKING WITHIN THE MICROBUB-
BLE'S SURFACTANT MONOLAYER

The above-mentioned artificial microbubble surfactant mixtures, and other successful mixtures found for stable microbubble
production (ref. 544-546), all contain saturated glycerides (with
acyl chain lengths greater than 10 carbons) combined with
cholesterol and cholesterol derivatives (cf. Chapters 9 and 10, and
ref. 544). As described earlier, long chain lengths in nonionic (or
even unionized) surfactants are known to favor the formation of
both large, rodlike micelles (as opposed to small spherical
micelles) and macroemulsions (as opposed to microemulsions) (see

Sections 10.2 and 9.1, respectively). Hence, the observed generation of a concentrated gas-in-liquid (macro)emulsion which interacts readily with simultaneously formed large (rodlike) micelles, using the above surfactant mixtures, is to be expected from and confirms such molecular packing considerations.

Similarly, the cholesterol and cholesterol derivatives present in the surfactant mixtures employed also favor the formation of coarse (gas-in-liquid) emulsions (cf. ref. 454). This trend arises from the well-known condensing effect (ref. 547-551) of cholesterol or cholesterol-like steroids on long-chain glycerides (of many types examined) in monolayers at the gas/water interface (at 21°C). Evidently, the interfacial mixed monolayer is too highly condensed with the strongly associating components to permit any great degree of curvature (ref. 454) and prevents very small (microemulsion) droplets, i.e., ultrasmall microbubbles (< 0.05 μm) in this case, from being produced.

An additional factor may also contribute to the tight molecular packing in the microbubble's surfactant monolayer which, again, opposes strong curvature and minimizes gas leakage from the microbubble. Specifically, while the saturated glycerides contained in the artificial microbubble surfactant mixtures employed to form artificial microbubbles were predominantly monoglycerides (see Chapters 9 and 10), up to 25 mole % of the glycerides represented triglycerides (ref. 544). This saturated triglyceride content may be significant since it has been suggested that in a monolayer of mixed saturated glycerides, the hydrophilic regions of monoglyceride versus diglyceride or triglyceride molecules, even with the same acyl chain length, may not be at the same level at an interface (ref. 552). This would facilitate closer packing of the different types of saturated glyceride molecules when they associate in the monolayer (ref. 552-554).

11.3 REPULSIVE HEAD-GROUP INTERACTIONS AND MONOLAYER CURVATURE

Still another factor favoring the formation of coarse gas-in-liquid emulsions, as demonstrated in Chapter 10, is that all the surfactants employed were nonionic. This feature results in weaker repulsive interactions among the (polar, but uncharged) head groups in the surfactant monolayer surrounding each artificial microbubble. Accordingly, the forces responsible for curvature in the surfactant monolayer are weaker (cf. ref. 555) and, hence, larger microbubbles are favored. (The same physical interrelationships contribute to the finding that with nonionic (and/or unionized) surfactants, the formation of large rodlike micelles is favored over small spherical micelles (see Section 10.2 and ref. 555).)

However, the various above-described factors (i.e., uncharged head groups, monoglyceride-diglyceride/triglyceride association, cholesterol condensing effect, and long hydrocarbon chain length) favoring larger microbubbles (and larger micelles) are, in fact, eventually fully opposed by a repulsive force which must come primarily from the surfactant head groups (cf. ref. 450,556). In a multimolecular aggregate formed by surfactants with uncharged head groups (such as the hydroxyl-containing glycerol moiety of glycerides or the nondissociable organic acid moiety of cholesterol esters), a preference for hydration (ref. 557), as opposed to self-association, is involved (ref. 450). This (noncoulombic) head-group repulsion increases as head groups are forced closer together by an increase in microbubble size and, hence, growth of the microbubble eventually comes to a halt.

11.4 MICROBUBBLE FISSION, COLLAPSE, AND RE-EMERGENCE

Along with the greater head-group repulsion with increased size of a microbubble, there is an increased likelihood for major distortion of the spherical shape of the enveloping surfactant monolayer by collision of the microbubble with surrounding microbubbles. Such distortion can lead to significant local alterations in head-group spacing and resultant instability in the microbubble's surfactant monolayer.

This molecular argument may explain the earlier-described and repeatedly observed finding with concentrated gas-in-liquid emulsions (see Section 10.4) that, following a period of microbubble growth, the average hydrodynamic diameters (detected by photon correlation spectroscopy) of the microbubble and micellar populations simultaneously decreased. In such situations, microbubble fission may have occurred (following microbubble collision) and the surfactant molecules needed for the expanding

202

gas/water interface accordingly were obtained from the rodlike micelles. This explanation is consistent with the fact that the proportion of micelles is seen to decrease relative to the microbubble population as microbubble diameter decreases (see Section 10.4). It is also consistent with the independent, earlier experimental observation that gas microbubble fission readily occurs, upon microbubble collision, in some physiological and other solutions where high concentrations of surfactants are present (ref. 521), i.e., well above the critical micelle concentration.

Following microbubble fission within the concentrated gas-in-liquid emulsions discussed in Chapter 10, the dynamic light scattering data indicate that the ultramicrobubbles produced probably shrink down and form the large micelles. This ultramicrobubble-to-micelle progression is to be expected since the preceding process of microbubble fission into ultramicrobubbles results in higher curvature and hence greater gas permeability (cf. Section 11.2) of the enveloping surfactant monolayer, along with a higher pressure inside the ultrasmall microbubbles (due to increased Laplace forces), all of which act to bring about the loss of gas from the ultramicrobubble.

Once collapse into a rodlike micelle has occurred, various physicochemical factors (cf. Section 10.3) contribute to the re-emergence of a surfactant-stabilized microbubble. Firstly, the very large rodlike micelles (up to at least 380 nm in [end-to-end] length) formed by nonionic surfactants, as reviewed in Section 10.2, would each be expected to solubilize a relatively large quantity of gas compared to an ordinary, small spherical micelle. The same large gas-solubilization capacity is to be expected of any large, rodlike micelles formed from the nonionic surfactants comprising the earlier-described surfactant mixtures used to produce artificial gas microbubbles (see Chapters 9 and 10). Secondly, the gas-rich interior of these probable large, rodlike micelles of artificial microbubble surfactant might well reach a state of supersaturation if, during the flexing motions characteristic of large rodlike micelles (ref. 466,479), appreciable water penetration into the micelle were to occur (cf. ref. 519,520). (Such penetration appears likely in view of evidence accumulated over the last two decades (ref. 459,519,520,558-561; cf. ref. 562) pointing to the existence of some bound water in the interior region of even small spherical micelles, which lack the flexing motions of rodlike micelles, near the first few methylene groups adjacent to the hydrophilic head groups. It is therefore useful to divide the micelle interior into an outer core which may be penetrated by water molecules and an inner core from which water is excluded (ref. 459; see also ref. 562).) Following water penetration, the combined factors of gas supersaturation (and hence impending phase separation), a low surface tension, and an enveloping surfactant monolayer (which, in turn, is surrounded by the saturated, aqueous surfactant solution) together could reasonably be expected to lead to microbubble formation.

In Chapter 10, particle size distributions, derived from photon correlation spectroscopy of artificial microbubble surfactant solutions, were presented which did in fact indicate the formation of lipid-coated microbubbles from large micellar structures (see also Chapter 15). Moreover, direct optical imaging of (at least the largest diameter) lipid-coated microbubbles formed from such Filmix surfactant solutions was subsequently accomplished (see Fig. 11.1) in a separate study, using phase-measurement interferometric microscopy (ref. 527). This latter study also included microbubble size distribution analyses (in this case based on electroimpedance-sensed volumetric sizing), at different salt concentrations and after the addition of thiourea, which clearly indicated that electrostatic interactions are much less important than hydrogen bonding in stabilizing the (nonionic-)lipid monolayer surrounding each microbubble (ref. 527). This particular "finding parallels the physicochemical stabilization mechanism operating within naturally occurring surfactant-stabilized microbubbles found in various surface waters; specifically, the surfactants in the [natural] microbubble's monolayer coating are reversibly held together by hydrogen bonding and nonpolar interactions" (ref. 527) (see also ref. 356,361, and Chapters 5 and 6).

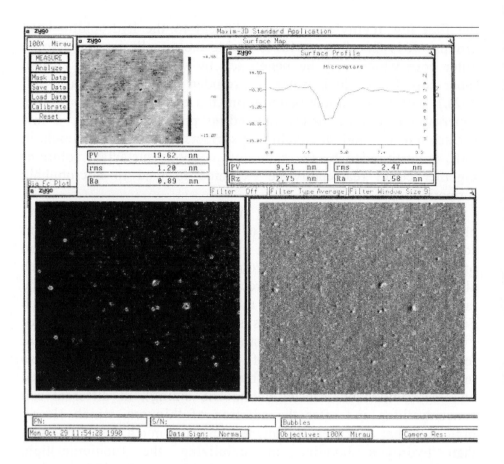

Fig. 11.1. Direct optical imaging of lipid-coated microbubbles (LCM) by phase-measurement interferometric microscopy (see text). (Taken from ref. 527.)

Chapter 12

TARGETED IMAGING OF TUMORS, AND TARGETED CAVITATION THERAPY, WITH LIPID-COATED MICRO-BUBBLES (L.C.M.)

The Filmix surfactant-stabilized microbubbles described in Chapters 9-11, also referred to as concentrated gas-in-liquid emulsions or lipid-coated microbubbles (LCM), have been modeled from natural microbubble surfactant (cf. Chapters 6 and 7) and selected physiological surfactants (cf. Sections 8.5 and 9.1). Accordingly, Filmix coated microbubbles or LCM contain specifically nonionic lipids exclusively throughout their microbubble coating. In numerous published animal studies, outlined in the next two chapters, it has been repeatedly observed that the nonionic-lipid monolayer coating of LCM causes this specific synthetic-microbubble agent to display marked tumor-targeting abilities well-suited for both diagnosis and treatment of solid tumors.

12.1   DESCRIPTION OF THE L.C.M. AGENT (FILMIX®)

The patented Filmix® agent is a stable suspension of air-filled, lipid-coated microbubbles in isotonic saline (i.e., 0.9% w/v NaCl in distilled water). The proprietary surfactant mixture, used to form each microbubble's monolayer coating, contains only low-molecular-weight nonionic lipids -- comprising saturated glycerides (with acyl chain lengths greater than 10 carbons) combined with cholesterol and cholesterol esters (ref. 538-546). This powdered surfactant mixture is added to the isotonic saline to a level (20 mg/100 ml) where the total surfactant concentration is well above the critical micelle concentration needed to form any mixed micelles, but is below the point of significant phase separation; the resultant saturated solution is shaken vigorously (mechanically, i.e., no sonication necessary), and subsequently filtered through a 0.45-μm pore-diameter membrane filter. The aqueous Filmix®

suspension is stable in vitro for more than one year when stored refrigerated (ref. 524,525,536). These lipid-coated microbubbles also have the important qualities of uniformity of size and extremely small diameters (see below).

Quality-assurance testing of the LCM agent has included particle size analysis, by electroimpedance-sensed volumetric sizing using a Coulter Multisizer, which consistently yielded the following results: Using a Coulter aperture tube with a 50-μm orifice (giving the instrument a nominal particle-diameter detection range from approximately 1.0 μm to 30.0 μm), more than 99% of the detected microbubble population is under 4.5 μm, and all microbubbles are less than 5.0 μm in diameter (ref. 524). In addition, one-by-one nonoptical counting of particles (i.e., those having diameters *larger* than ~ 1.0 μm) by the Coulter Multisizer consistently resulted in a calculated total concentration of approximately 5 x 10^5 microbubbles/ml in the Filmix® samples (ref. 525,536). All of the above product specifications remain constant for many months when the LCM agent is stored at room temperature, and for over one year when the LCM agent is stored refrigerated (but not frozen) (ref. 524). Furthermore, when the entire particle-size analysis (Q/A testing) was repeated on many Filmix® batches using subsequent, more sensitive models of the Coulter Multisizer (Models II and IIe), maximum microbubble diameter remained under 5.0 μm. However, these newer Coulter instruments uncovered evidence that the *vast majority* of the Filmix® "microbubble/particle" population exhibits diameters *less* than 1.0 μm (see below for further discussion; see also Chapter 15).

The lipid-coated microbubbles in Filmix® agent have never been found to agglomerate nor coalesce into any "microbubble/particle" structure larger than 5 μm, either in vitro or in vivo, thus the risk of air embolus is negligible (ref. 524). Acute intravenous toxicity studies of this (*isotonic*) LCM agent in rabbits and dogs were conducted at an independent GLP contractor (ref. 563,564). The acute intravenous LD_{50} in both species was found to be greater than 4.8 ml/kg. Furthermore, no signs of gross toxicity or mortality were observed at a dosage of 4.8 ml/kg (ref. 563,564). It has also been found in other animal toxicology studies that at intravenous Filmix® doses of 0.14 ml/kg, given three times per week for six weeks in rats, and 0.48 ml/kg given three times per week for three months in rabbits, there were no untoward changes in serum chemistry, liver functions, hematology or clotting profile or histological changes in adrenals, bladder, brain, heart, kidney, liver, lungs, marrow, pituitary, spleen, testes, thyroid or ureters (ref. 536).

12.2 TARGETED ULTRASONIC IMAGING OF TUMORS WITH L.C.M. AS A CONTRAST AGENT

Due to the small size of these lipid-coated microbubbles (all below 5 μm, and vast majority well under 1.0 μm), they were expected to cross the pathologic fenestrations known to exist in intrinsic tumor capillaries (ref. 565-568). (These capillary fenestrations explain the long-known "leakiness" of tumor capillaries, which in the case of brain tumors includes breaches in the blood-brain barrier (ref. 566,568).) The first tumor study using Filmix® (ref. 525) demonstrated that LCM, injected intravenously, did in fact cross into, and *intensified* the sonographic images of, gliomas in the rat brain. As displayed in Fig. 12.1, this *echogenic enhancement* of tumors typically peaked within only a few minutes after contrast injection (Fig. 12.1, top right). The enhancement persisted continuously for over 30 min after the LCM agent dose was administered (0.15 ml/kg in all animals). (Fig. 12.1 also shows a schematic representation of the tumor generated from intensity data (left bottom), and a photograph of an axial section of the cerebral tumor stained with hematoxylin and eosin (right bottom).) Furthermore, when the LCM agent was injected (intravenously) once per day in rats, it was found that initial (ultrasonic) detection of developing brain tumors (C6 gliomas) averaged only 4.09 ± 0.08 days (mean $\pm$ SEM; n = 11) after tumor innoculation versus 6.67 ± 0.35 days (n = 9) for controls ($P <$ 0.001). Hence, sonographic detection of gliomas occurred approximately 40% (i.e., 38.7%) earlier with the use of LCM contrast medium, intravenously injected in the rat (ref. 525). All brain tumors identified by ultrasound scans in this study were later confirmed by histological examination (ref. 525,526).

In a subsequent brain-tumor study (ref. 526) using LCM along with a lipid-specific stain (cf. 569), a detailed evaluation of the actual distribution of lipid-coated microbubbles in the tumor and surrounding organ was conducted and compared to the distribution of echoes (from the microbubbles) on the sonogram.

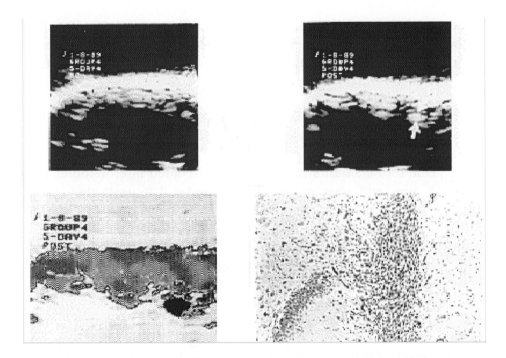

Fig. 12.1. These coronal scans are representative ultrasound images of a rat cerebral glioma before (top left) and less than 2 minutes after intravenous injection of 0.15 ml/kg of lipid-coated microbubbles (top right). A schematic representation (lower left) of the tumor (shaded) is presented in the context of the coronal scan. This image has been created from the actual map of the pixel intensities of the scan. The results are presented with black and white reversal for ease of interpretation. Lastly, a photomicrograph of the same cerebral tumor site, sectioned in the axial direction, is shown (lower right). (Taken from ref. 525.)

Histological preparations were used to verify the location of the LCM in the tumor, as presented in Fig. 12.2. (In each of the histological sections prepared from a given tumor, a needle track, as well as an injection/expansion artifact, left a void not only in the H&E sections, but also in the LCM aggregation. This void was not tumor necrosis; the tumor has no higher grade than anaplastic astrocytoma, and in most areas was astrocytoma. By counting the stained LCM in each 100 x 100 μm square, data were assembled for contour mapping. Counts ranged from 0 (in the void) to 17/square

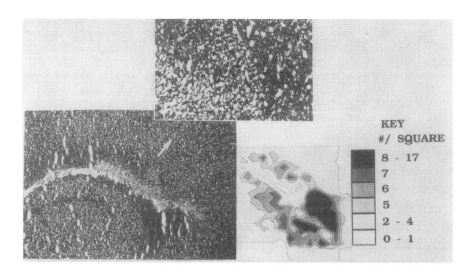

Fig. 12.2. Microbubble count contour map. Lipid-coated microbubble (LCM) distribution in the tumor is represented by lines of microbubble isodensity in this contour map (bottom right). Notice that the area of needle/expansion artifact seen in the photomicrograph (bottom left) corresponds to a nil microbubble isodensity. The exploded panel from the photomicrograph clearly demonstrates the malignant features of the glioma. There is a significant freezing artifact in the left side of the photograph. (Taken from ref. 526.)

(in the pleomorphic part of the tumor).) As shown in Fig. 12.2, there is a correspondence between the histologically defined tumor and high LCM concentration. The contour map of lipid-coated microbubble density shows the correspondence of the microbubble void to the injection artifact. In this figure, by counting all squares with an LCM density greater than four per square (not including the void containing no LCM), the tumor area was determined to be 1.50 mm^2. If LCM densities greater than five per square were considered, the LCM-specific area was 0.96 mm^2. At an LCM density of greater than six per square, the LCM-specific area was 0.57 mm^2. By histological examination, the area of the tumor in an adjacent slice 8 μm away was determined to be 0.93 mm^2. The number of LCM outside the tumor decreases rapidly with distance; the background LCM density beyond the immediate tumor (i.e., rat-brain glioma) region and in the contralateral hemisphere was one or two per square. The data from three other rats were similar; the LCM were found concentrated in the tumor, sparse in the normal brain, but were not found in void areas (ref. 526).

In this same study, a similar centrifugal attenuation of echo intensities was noted in the contour analysis of intensity, particularly when the LCM contrast agent was employed. The analysis was primarily of qualitative value in demonstrating the spread of intensities due to LCM contrast agent. (As concerns the methodology of the echo-intensity analysis, the ultrasonographer was not blinded to the presence or absence of LCM. This experimental bias was mitigated by two procedural points. First, the arrangement of the transducer and the relative sizes of the transducer and the brain guaranteed that the tumor-containing sector was scanned, and that actual tumors were consistently in the same area, guaranteed by the stereotaxic placement. Thus, the location of the window was nearly constant. More important, the computer was blinded to the experimental conditions. The signal-to-noise ratio or the two-dimensional power spectrum was computed based solely on the contents of the window; the presence or absence of LCM was totally transparent to the computation.) In summary, the LCM go directly to the tumor and almost uniquely to the tumor, with little accumulation in the adjacent normal brain. They remain there for at least 30 minutes. The allocation of LCM was verified by histologic examination, and their distribution pattern had likewise been shown. On the scan itself, the LCM contrast permitted visual and mathematical identification of low-grade tumor. The changes in echogenicity caused by the LCM contrast could be described in a number of ways: by simple distribution of intensity, by signal-to-noise ratio, or by texture. Distribution of intensity presumably was governed by the distribution of the LCM themselves, as the microbubble-lipid stains suggested.

In still other, subsequent, contrast-assisted ultrasonic imaging studies utilizing different tumor types, different organs, and/or different animal species, it was again found that intravenous injection of the LCM agent was effective in targeting both experimental malignant subcutaneous tumors (carcinosarcomas) and liver tumors in the rat, as well as spontaneous malignant tumors of the prostate, liver, and spleen in dogs (ref. 570). In addition, in each study, intravenously injected LCM again targeted the tumors specifically, noticeably enhancing the signal-to-noise ratio in the lesion without enhancing any of the other anatomic structures in the abdominal cavity and the pelvic area (ref. 528,570). In addition, lipid-specific stains indicated that the LCM lodged selectively within the subcutaneous carcinosarcomas and liver tumors of the rat, but not in the adjacent soft tissue (ref. 528).

The above liver-tumor study (ref. 528), in particular, included some other features of interest. Specifically, an entirely different lipid-specific stain, "Oil Red-O" (ref. 571), was used in this case (as opposed to the earlier brain-tumor studies) to verify LCM distribution in the liver tumors (i.e., Novikoff hepatomas), and in this particular case the ultrasound transducer could be kept immobolized (in a stand) during the entire scanning session. (As explained in the study report: "The transducer was secured at a fixed angle, and the animals [i.e., rats] remained in the same position (under general anesthesia) during the experiment to ensure that the images acquired at each time interval scanned the same volume segment of the liver (encompassing both tumor tissue and normal liver parenchyma)" (ref. 528).) Fig. 12.3 is a representative series of rat liver sonograms, before and after injection of LCM, which shows the time course of sonographic enhancement by LCM of Novikofff hepatoma in anesthetized rats using an

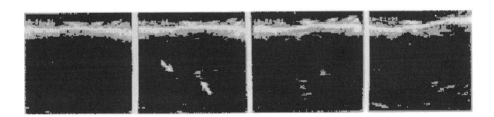

Fig. 12.3. Time course of sonographic enhancement by LCM of Novikoff hepatoma in anesthetized rats using an immobilized transducer. These representative rat liver scans were obtained before injection, and 2, 30, and 60 minutes after i.v. injection, respectively. Arrows indicate the relatively bright, contrast-enhanced tumor area in the liver (within the image plane) at 2 minutes after injection. Contrariwise, the same area before injection (first panel) is only slightly hyperechoic compared with the surrounding normal liver parenchyma. (Taken from ref. 528.)

212

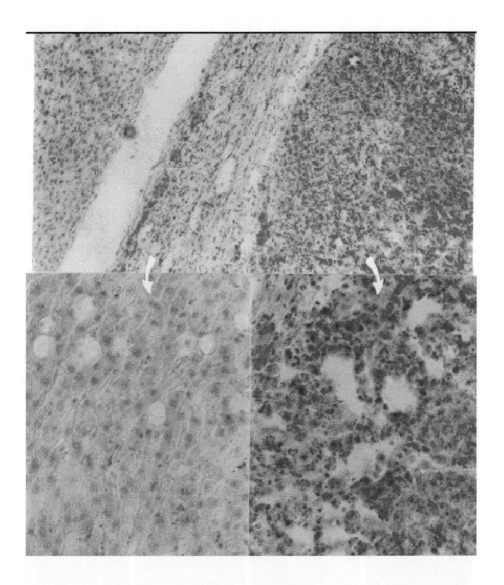

Fig. 12.4. Demonstration of tumor targeting ability of LCM after i.v. injection into a rat bearing Novikoff hepatoma. All histologic sections were stained with Oil Red-O and counterstained with hematoxylin. (Top panel) A low-power view of the hepatoma and surrounding normal liver tissue. (Bottom panels) High-power insets of the neighboring normal liver parenchyma (bottom left) and the Novikoff hepatoma itself (bottom right). The lipid-coated microbubbles can be appreciated as solid black discs ranging in size from submicron up to 4 or 5 μm. (Taken from ref. 528.)

immobilized transducer. These rat liver scans were obtained before injection, and at 2, 30, and 60 minutes after tail-vein injection, respectively, of the animal (with 0.2 ml/kg of LCM agent). It is evident from the figure that injection of the LCM agent produces a rapid increase (i.e., within ~ 2 min) in the echogenicity of malignant liver tumors in the rat; this rapid enhancement in the signal-to-noise ratio of the Novikoff hepatoma was not accompanied by other enhancement of any surrounding anatomic structures in the abdominal cavity (ref. 528). Upon calculation of the mean persistence of tumor image enhancement over a 1-hour period after injection, a statistically significant difference in the signal-to-noise ratio was observed between the group that received LCM (10 rats) and the group that received saline (6 rats) ($P < .01$). The visual effect persisted for 30 minutes after contrast injection. Further support for the tumor-targeting ability of LCM can be found in Fig. 12.4; it is a composite photograph of liver tumor histology with high-power insets of hepatoma versus surrounding normal liver parenchyma, all counterstained (with Oil Red-O) for the presence of LCM. In the figure, the lipid-coated microbubbles can be appreciated as solid black discs ranging in size from submicron to 4 or 5 microns (ref. 528); from examining this micrograph and similar ones, the ratio of tumor-to-tissue LCM number density has been measured to be approximately 1300:1 for these liver tumors (ref. 530).

12.3 TUMOR DETECTION VERSUS TUMOR THERAPY WITH L.C.M.

While the highly selective targeting and rapid accumulation of LCM into different solid tumors is quite marked and ought to be exploited clinically, there is still considerable room for improvement in the actual signal-to-noise ratio of the tumor image enhancement on ultrasonograms arising from the LCM accumulated within the tumor mass. This particular adjustment in the imaging characteristics of LCM would be desirable for more effective tumor detection and rapid screening, but is not necessary in the separate case of conducting tumor *therapy* with LCM (see below).

Considering in the first instance only diagnostic imaging and screening for tumors, one solution could be to use an entirely different imaging modality with better resolution such as magnetic resonance imaging, where the targeted LCM agent still serves as an appropriate contrast agent (see section 12.4). Another solution could be to adapt ultrasound equipment for optimal use with the targeted LCM agent, e.g., by exploiting the nonlinear properties of lipid-coated microbubbles (cf. ref. 572). This second type of approach is currently being used by various investigators (e.g., ref. 572-574) to examine a variety of other microbubble contrast agents: For example, Krishna et al. (ref. 574) explain that the nonlinear behavior of contrast agents in an acoustic field leads to harmonic components in the backscattered signal; specifically, they showed that microbubbles with various surface coatings generate significant subharmonics under various insonating conditions. In addition, Forsberg et al. (ref. 572) argue that ultrasound contrast agents promise to improve the sensitivity and specificity of diagnostic ultrasound imaging, since given the lack of subharmonic generation in tissue, one approach is the creation of subharmonic images by transmitting at the fundamental frequency (f_o) and receiving at the subharmonic ($f_o/2$). Such imaging should display a much better lateral resolution and may be suitable for scanning deep lesions due to the higher transmit frequency and the much smaller attenuation of scattered subharmonic signals (ref. 572).

Interestingly, besides probably improving diagnostic imaging of tumors, subharmonic emission can very likely also be used to effectively monitor the ultrasonic treatment and ultrasonically-induced destruction of tumors (see below). For example, Morton et. al. describe work in which subharmonic emissions from ultrasonically irradiated biological samples are integrated over time, and the resultant signal (which is believed to be indicative of cavitation activity) is found to correlate well with the extent of cellular damage. These authors go on to discuss the potential of using subharmonic emission monitoring as a quantitative predictor of ultrasonically induced biological damage, both in vitro and in vivo (ref. 575). Similarly, Jeffers et al. more recently have reported that detection of subharmonics confirmed the presence of cavitation, and that cell lysis was well correlated with the subharmonic amplitude. These authors conclude the results show that albumin-stabilized microbubbles, similar to those currently used commercially as ultrasound contrast agents, may provide a significant source of nuclei and improve prospects for cancer therapy using acoustic cavitation (ref. 576). In summary, subharmonic emission monitoring has been used by these investigators as a quantitative predictor of ultrasonically induced lysis or cavitation-related destruction, readily facilitated by the local presence of coated microbubbles (acting as gas nuclei), of identified human cancerous cells. This combined diagnostic/therapeutic finding fits in well with, and should further facilitate, the earlier-mentioned planned additional use of the LCM agent for ultrasonic therapy of LCM-targeted tumors. In fact, a preliminary cavitation-therapy study has been published on LCM-facilitated ultrasonic treatment of experimental malignant tumors in rats (ref. 530). It was found that when the animal tumors were insonated (with continuous-wave ultrasound [0.25 W/cm^2]) after injected LCM had targeted and accumulated within the tumor mass, selective necrosis, lymphocyte proliferation, and hemorrhage within the tumor could be demonstrated (see Section 12.5 for further discussion). In the future, with adjustments in the ultrasound equipment and subharmonic emission monitoring of a given tumor site, noninvasive LCM-facilitated ultrasonic therapy of liver tumors should move closer to becoming a clinical reality (see Section 12.5).

12.4   USE OF L.C.M. AS A TARGETED, SUSCEPTIBILITY-BASED, M.R.I. CONTRAST AGENT FOR TUMORS

Ultrasonographic-assisted approaches to various lesions within the human nervous system are now commonplace in neurosurgical operating rooms. The presence of bone (i.e., skull or spine) has been a technical constraint of ultrasonography in all but pediatric cases in which the fontanels had not closed. Because of the development of smaller transducers, it has now been possible to use ultrasonography through small trephine openings, through craniotomy flaps, and through laminectomy incisions to image the central nervous system.

The potential for intra- and postoperative ultrasonographic monitoring of deep brain tumors with the targeted LCM agent suggests, in turn, the added possibility of preoperative diagnosis and localization of brain tumors with magnetic resonance imaging (MRI), which is not limited by the presence of bone (an intact skull). As it turns out, the above-documented ability of LCM to concentrate rapidly and selectively in tumor tissue (cf. Sections 12.1-12.3), with no accumulation in surrounding normal tissue, has been found (ref. 529) to cause a significant change in the "bulk magnetic susceptibility" (BMS) inside tumors in vivo as detected by MRI (see below).

Since the BMS of air is about 0.4 ppm (37 °C) (ref. 577) and that of tissue or blood is about -9 ppm (ref. 577), the susceptibility difference created by the accumulation of LCM in the tumor region was expected to generate local magnetic field gradients that shorten the average T_2 and/or T_2* values of the water proton spins inside the tumor. Therefore, the affected region should appear darker in a T_2- or T_2*-weighted MR image, with the larger contrast effect in the latter (ref. 529). MRI contrast enhancement by the LCM agent was examined in a rat brain tumor model (9L gliosarcoma) and the time courses of the enhancement were recorded.

The MRI experiments were performed on a clinical 1.5 T GE SIGNA Advantage Unit with the combination of a body transmitter coil and a home-made 4-cm diameter solenoid receiver coil that was placed around the rat's head. The rat's body axis was perpendicular to the magnetic field. A 3D volume data set was acquired using a 3D GRASS (Gradient Recalled Acquisition in Steady State) sequence in about 8 min, with a 256 x 256 data matrix size, 8 cm FOV, 5° flip angle, TE = 15 ms, and TR = 60ms. A total of 28 contiguous slices with 0.7 mm thickness were collected (ref. 529).

Prior to MRI examination, the tumor-bearing rat was anesthetized (i.p.) and the tail vein was catheterized with a 27-gauge lymphangiogram needle. A dose of the LCM agent (Filmix®, 0.3 to 1.4 ml/kg) was delivered by intravenous injection through the tail vein over 1 to 2 min. One 3D volume GRASS data set was acquired before and another at about 3 to 5 min after the injection of LCM agent. Rats were imaged 7, 14, and 19 days after the tumor cell implantation. A dose (0.5 mmol/kg) of a standard clinical MRI contrast agent (Magnevist [Berlex]) stock solution (500 mM GdDTPA) was later delivered through the tail vein of the rat with a 19-day tumor to also define the tumor region in a T_1-weighted image using a spoiled-GRASS sequence (ref. 529).

The pre-LCM (i.e., pre-Filmix®) GRASS images showed no delineation of the 7- and 14-day tumors. The post-LCM T_2*-weighted GRASS images showed one or several dark focal regions (area ranging from 0.11 to 1.29 mm^2) at the location of the tumor, but not elsewhere in the normal tissues within the field of view.

The older the tumor, the more abundant and larger the focal regions. The T_1-weighted post-Magnevist image of the rat with a 19-day tumor exhibited a much larger ($\sim$ 23 mm^2) bright region in the brain that is presumably the area of pathologically altered blood-brain barrier. All the dark regions in the T_2*-weighted post-LCM image were located within the Magnevist-enhanced area (ref. 529).

Since the volume fraction of the total LCM in the Filmix® suspension is roughly 2-3 x 10^{-6}, the BMS of the Filmix® stock solution is almost the same as that of pure saline solution, about -9 ppm (ref. 577). Thus, the unconcentrated stock solution would not cause BMS-based contrast in tissue. The results suggest that the LCM are indeed concentrated directly and uniquely in the tumor, with no accumulation in normal brain tissue (ref. 529). This is probably due, in part, to pathological alterations in the intrinsic tumor capillaries (ref. 565-568) (cf. Section 12.2), and appears to significantly increase the local volume fraction of the microbubbles and therefore the local BMS value. The change of the BMS in the tumor region finally results in contrast enhancement in T_2- and T_2*-weighted images. The regions appearing very dark on the GRASS images must have the most concentrated LCM inside the tumor, and possibly inside tumor cells themselves (ref. 529).

12.5 L.C.M.-FACILITATED ULTRASONIC THERAPY OF TUMORS

Sections 12.1-12.4 have reviewed how the highly selective targeting and rapid accumulation of LCM into different solid tumors has been utilized for enhanced diagnostic ultrasound and magnetic resonance imaging of tumors. This repeated observation (see also Chapter 13) that LCM are heavily concentrated in various tumors, with no accumulation in surrounding normal tissue, led to identification of another important aspect of LCM: potential use as a targeted therapeutic agent. The degree of targeting possible with LCM depended on the tumor type; for example, in subcutaneous tumors (Walker-256 carcinosarcoma) the ratio of tumor-to-tissue LCM number density was counted to be 700:1; in liver tumors (Novikoff hepatoma), the ratio was counted to be 1300:1 (ref. 530).

The first mode of targeted tumor therapy explored, using intravenously injected LCM, involved simply a stronger vibrating and/or bursting of these coated microbubbles by employing higher diagnostic-ultrasound output intensities (i.e., continuous-wave ultrasound). Such LCM-facilitated "ultrasonic tumor therapy" is based on the established fact (ref. 537 for literature review) that gas pockets, such as those provided by the LCM, are capable of acting as heat sinks as well as cavitation nuclei when exposed to some forms of diagnostic ultrasound (ref. 538-543,576).

This series of collaborative experiments began by examining LCM-facilitated cavitation in a (Walker-256 carcinosarcoma) brain metastasis model, insonating through a craniectomy in the rat's skull. We used moderate (continuous-wave Doppler) diagnostic doses, but noted a persistent capacity to injure neurons in deep cortex and severely injure hippocampal neurons (ref. 530). Seeking a better preliminary model, we abandoned brain-tumor therapy and used a subcutaneous model, where there were no evident heating effects on the adjacent (embedding) tissues.

Specifically, we studied the LCM-facilitated heating/cavitation efffects on a subcutaneous Walker-256 tumor in the rat (ref. 530). In brief, after tumor implantation, rats were divided into 3 treatment groups to be evaluated at 4 time intervals after treatment (0, 24, 48, and 72 hours). All scans were made with a 7.5 MHz duplex probe using an aqueous gel standoff (2 cm x 9cm). The transducer was secured at a fixed angle during the experiment. Upon detection of the tumor with the mechanical scanner mode, the probe was switched to the continuous-wave (CW) mode (4.5 MHz). The Doppler beam axis was adjusted to maintain a standard angle, and the tumors were continuously insonated for 8 minutes through the gelatin stand-off. The spatial average, temporal average (SATA) intensity for this experiment was 0.25 W/cm^2. Rats from group 2 received ultrasound treatment, but no LCM injection. For the sham-control rats (group 1), the probe was attached to the tumor area but no treatment was performed. The histology was reviewed by a neuropathologist made blind to the treatment applied to the individual specimens. The lesions were graded for necrosis; hemorrhage and lymphocytic infiltration were graded on relative scales. There was no change in the temperature, as a result of ultrasound treatment, of the two tumors measured within the sensitivity of our thermometer ($\pm 0.5°C$) (ref. 530).

Differences were found among the animals without insonation and in animals insonated, versus animals insonated after the intravenous injection of LCM. Effect was measured by degree of necrosis, lymphocyte proliferation, and hemorrhage in 16 animals studied. The effects were most pronounced 48 hours after treatment (ref. 530).

In summary, rats harboring subcutaneous carccinosarcomas were divided into three groups. Group 1 (control) received no treatments. The tumors of Group 2 were insonated for 8 min with ultrasound at frequencies and power within diagnostic range of the machine. Group 3 was similarly insonated just after the LCM were injected. Necrosis and hemorrhage occurred following insonation. In the presence of LCM, these changes were uniformly and unequivocally increased (ref. 530).

With regard to follow-up studies, although LCM-targeted hepatomas in the liver have not yet been subjected to ultrasonic tumor therapy, destruction of such tumors should be more easily accomplished with even less risk to the surrounding normal organ because of the documented (ref. 528,530) higher ratios of tumor-to-tissue LCM concentration. It is anticipated that noninvasive, ultrasonic tumor therapy of LCM-targeted liver tumors in humans, if eventually developed and commercialized in the future, would have substantial clinical utility (ref. 578). Malignant tumors arising in the liver, or metastasizing there from another site, are a difficult and relatively common clinical problem worldwide: Primary liver cancer has an annual incidence of at least 1 million cases worldwide (ref. 579,580); liver metastases occur in over 40% of cases of metastatic cancer and is the most common cause of death in all cancer patients (ref. 579,581,582). Effective management options are limited, even at modern tertiary care centers (ref. 578).

Chapter 13

TARGETED DRUG-DELIVERY THERAPY OF TUMORS USING
L.C.M.

13.1   INTERNALIZATION OF L.C.M. BY TUMOR CELLS IN
VIVO AND IN VITRO

In Chapter 12, it was reviewed how the highly selective
targeting and rapid accumuulation of LCM into different tumors
enhanced ultrasonographic and magnetic resonance imaging of the
tumor mass, and offered the additional prospect for a means of
accomplishing noninvasive selective treatment and/or destruction
of deep-seated tumor targets, in vivo, by using LCM-facilitated
ultrasonic tumor therapy. These same targeting attributes of the
LCM makes it a suitable candidate for selective drug delivery to
solid tumors, especially if it could be shown that LCM are actually
internalized by the tumor cells themselves.

The following section describes a detailed analysis (ref.
531) of the interactions of LCM with C6 glioma and 9L gliosarcoma brain tumors in the rat, and with the same tumor cells grown
in monolayer culture. The distribution of LCM in the brain parenchyma versus within the tumor was assessed in two ways. First,
unlabeled LCM were administered to rats and their presence was
visualized after sectioning the brain and staining with Oil Red-O, a
dye that recognizes cholesterol esters (ref. 571) in the LCM
coating.   Second, LCM were labeled with a fluorescent marker
prior to their intravenous injection into the animals. The dye, 3,3'-
dioctadecyloxacarbocyanine perchlorate ("diO"), has been used in
other studies to trace live cells (ref. 583,584) and because of its
lipophilic nature could be incorporated into the LCM without
changing their properties. DiO-labeled LCM were visualized
directly by fluorescence microscopy after fixing and sectioning the
brain. (LCM and diO-labeled LCM were kept at room temperature, in the dark, and used within 20 min of preparation.) The

interactions of diO-labeled LCM with tumor cells were also analyzed at the cellular level by confocal laser scanning microscopy. Previous work (ref. 585-587) has shown that this technique can resolve intricate cellular morphology and the localization of organelles within cells. Consequently, the interactions of diO-labeled LCM with tumor cells were examined at the cellular level by this technique, both in brain tissue and in cells in culture, in order to elucidate the mechanism of LCM uptake (ref. 531).

13.1.1 LCM reach tumors within minutes after i.v. injection: Light- and fluorescence-microscopy data

In order to detect the presence of LCM in the brain tissue, the Oil Red-O staining technique was employed. This method is only semi-quantitative, but useful because it does not require modifying LCM in any way prior to intravenous injection in animals. Rats bearing 9L gliosarcoma tumors in the brain were injected with the LCM agent in the tail vein (0.2 ml/kg) and sacrificed 2 min later. The brain was processed as described elsewhere (ref. 531). Light microscopy examination of brain slices containing 9L gliosarcoma tumor showed a tumor mass (Fig. 13.1, top left panel) recognized by the darker staining of the tumor cells' nucleus, surrounded by normal brain cells (to the right in that panel). Oil Red-O granules (black discs), indicative of the presence of LCM, are clustered over tumor cells, and absent from the surrounding parenchyma. In both C6 gliomas (Fig. 13.1, top right panel) and 9L gliosarcoma tumors the presence of LCM, as revealed by Oil Red-O staining, was always detected in the core of the tumor mass located deep within the brain, indicating that LCM reached the tumor within minutes of intravenous bolus administration (ref. 531). In some cases, rats bearing tumors were sacrificed at 30 min, 60 min, 6 hr, and 24 hr after the intravenous injection of LCM or diO-labeled LCM. The presence of LCM was detected by Oil Red-O staining, and that of diO-labeled LCM by fluorescence microscopy. The accumulation of LCM at the tumor site was maximal 30 minutes following the injection of LCM, but could still be detected after one hour. However, LCM could not be detected after 6 and 24 hours, which is in agreement with earlier ultrasound imaging studies indicating that LCM evanesce from the tumor site 1 hour after i.v. injection (ref. 531).

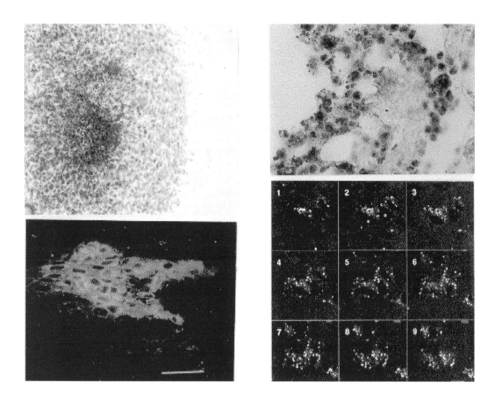

Fig. 13.1. This figure demonstrates confinement of the LCM to 9L gliosarcoma tumors in the Fischer rat (top left) and to C6 gliomas in the Sprague-Dawley rat (top right) with Oil Red-O lipid staining. Bottom left panel is the 9L tumor (in Fischer 344 rat) immediately after intravenous injection with diO-labeled LCM. Bottom right panel shows this same tumor. Here nine successive 0.8 μm tomographic sections in the Z axis from the confocal laser microscope illustrates the presence of the intracellular LCM. The scale bar is 10 μm. (Taken from ref. 531.)

13.1.2 <u>LCM preferentially interact with tumor cells in vivo: Data from confocal laser microscopy</u>

The distribution of LCM in the brain parenchyma was further analyzed using fluorescently labeled LCM and confocal laser scanning microscopy. As in the case of unlabeled LCM, rats bearing 9L gliosarcoma tumors were injected intravenously (i.e., via tail vein) with diO-LCM and sacrificed 2 min later. The brains were processed as described elsewhere (ref. 531). In this case,

224

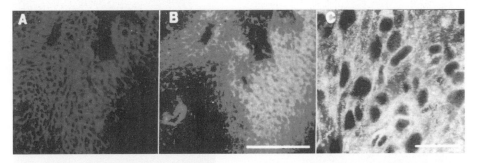

Fig. 13.2. Preferred association of LCM with tumor cells in vivo. Vibratome sections of brain of rats bearing 9L tumor that have been injected i.v. with diO-LCM, and sacrificed 2 min later. Sections in panels A and C were stained with TR-conjugated WGA. Specimens in panels B and C were observed by fluorescence microscopy using fluorescein optics. The scale bar represents 100 μm. (Taken from ref. 531.)

vibratome sections were counterstained with Texas Red conjugated wheat germ agglutinin (TR-WGA) to clearly display the tumor cells; WGA binds to tumor cells preferentially, and distinguishes the tumor area from the surrounding normal tissue. DiO-LCM were present in the solid central tumor (gliosarcoma) mass (Fig. 13.2(B)), and in several of the perivascular invasion clusters (Fig. 13.3(A)). Comparison of the area, in the same field of view, stained with TR-WGA (i.e., for tumor cells) (Figs. 13.2(A) and 13.3(B)) versus that stained with diO (i.e., containing diO-labeled LCM) (Figs. 13.2(B) and 13.3(A), respectively) indicated that LCM were associated with a large proportion of the tumor. Furthermore, the diO fluorescence pattern was always limited to the tumor area (ref. 531). This was also observed with rats bearing C6 glioma tumors (data not shown). DiO-LCM were not detected in normal surrounding tissue including microglia, which were identified by tomato lectin staining (ref. 588) and formed a ring around the tumors. Intravenous injection of diO alone, in animals bearing tumors, did not result in labeling of the tumors or any other parts of the brain, indicating that the ability to target tumor cells in vivo is the property of LCM and not that of the dye (ref. 531).

Longer intervals (up to 1 hr) between the time of diO-LCM injection and sacrificing of the animals did not result in labeling of

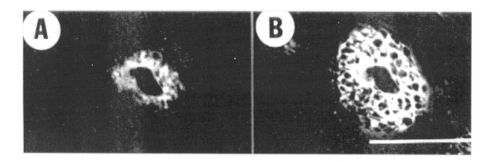

Fig. 13.3. This figure demonstrates the distribution of fluorescently tagged LCM in brain parenchyma analyzed by confocal laser scanning microscopy. Rats bearing 9L tumors were administered diO-LCM and sacrificed 2 minutes later. Vibratome sections were counterstained with TR-WGA which binds to tumor cells and distinguishes the tumor area from the surrounding normal tissue. Comparison of the area stained with TR-WGA (tumor cells) (B) and that stained with diO (A) indicates that LCM were associated with a large portion of the tumor. (Taken from ref. 531.)

normal brain tissue, suggesting that a preferential interaction of LCM with tumor cells is the basis for the labeling pattern obtained (ref.531).

13.1.3 <u>LCM are found inside tumor cells in vivo: Serial optical sections</u>

A high magnification view of the tumor (9L gliosarcoma) pictured above, in Fig, 13.2, revealed that LCM have been internalized by the tumor cells (Fig. 13.2(C)). The cell nucleus appears black (solid and oval); the diO-LCM are dark gray (with a speckled appearance), and immediately adjacent to the nucleus; the wheat germ agglutinin delineates the contours of the tumor cells, and overlap of diO-LCM with the agglutinin shows as bright areas. (The scale bar is 25 μm.) Serial optical sectioning confirmed the intracellular distribution of LCM (ref. 531) (see below).

Confocal laser scanning microscopy was used to determine the spatial relationship of LCM with tumor cells. Serial optical sections of 9L tumor cells situated in the brain parenchyma of rats having received diO-LCM were taken at 0.8 μm intervals. A series of nine consecutive sections was obtained; all nine of these are presented in Fig. 13.1 (bottom right panel). The nucleus of two cells, in panel section 2, is indicated by an "n", and the cytoplasm by a "c". The presence of labeled granules within the cells indicates that diO-LCM have been internalized by these cells (ref. 531).

13.1.4 LCM are endocytosed by tumor cells in culture: Kinetics of uptake and temperature dependence

In order to understand the interactions between LCM and tumor cells, the latter (C6 glioma) were grown in monolayer culture and exposed to diO-LCM in saline. Under the conditions used, the labeled LCM are spherical (i.e., each representing either a single LCM or possibly the agglomeration and/or coalescence of several smaller LCM) and have an "average diameter" of 0.5 μm ± 0.1, as measured by fluorescence confocal microscopy (ref. 531). C6 glioma cells incubated with diO alone, under the same conditions, constituted the control group.

The spatial relationship between LCM and C6 glioma cells was analyzed by fluorescence confocal microscopy at time intervals during incubation at room temperature (RT). After 5 min, the majority of LCM was still dispersed uniformly over the tumor cells and the areas of the substratum where no cells were present. However, after 30 min, there was an increase in the concentration of LCM on and/or in tumor cells, and after 60 min, the majority of LCM was associated with cells. A typical example of this association is shown in Fig. 13.4(A). The fluorescently labeled LCM are discrete entities, and do not appear to fuse with the tumor-cell plasma membrane. This was determined by measuring the size of LCM associated with the C6-glioma cells after 10 min incubation (at RT), a condition in which internalization is just beginning. It was found that 90% of the spherical LCM detected (which may each be a single LCM or possibly small agglomerations of several LCM, seen by fluorescence confocal microscopy) had a "mean diameter" of 0.5 μm. LCM remained as single spherical objects at the three temperatures under which LCM-cell interactions were monitored (4°C. RT, and 37°C). Isosurface rendering (with the confocal microscope's software package) of a portion of the tumor cell shown in Fig. 13.4(A) (cf. inset) confirmed that diO-labeled LCM are present initially as spherical, discrete objects (Fig. 13.4(B)) (ref. 531).

By contrast, the labeling of C6-glioma cells with the diO-

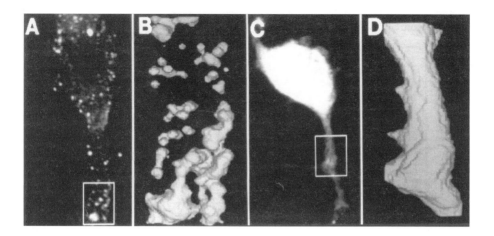

Fig. 13.4. Interactions of LCM with tumor cells in culture. C6 cells were incubated with diO-LCM (panels A,B) or diO alone (panels C,D) for 30 min at RT, and examined by fluorescence microscopy. The scale bar represents 10 μm. (Taken from ref. 531.)

dye alone (the control) is dramatically different, as seen in Fig. 13.4(C). In that case, a smooth pattern of uniform surface labeling was observed, suggesting that diO has become integrated with the membrane; in addition, isosurface rendering (cf. above) of a portion of the tumor cell (control) shown in Fig. 13.4(C) (cf. inset) confirmed the entirely uniform pattern of cell-surface labeling (Fig. 13.4(D)). (This finding is in agreement with results obtained with diO in other systems.) It indicates that the opposing <u>punctate</u> pattern observed with LCM (Fig. 13.4(A,B)) can only be obtained if diO remains associated with LCM during the incubation period (ref. 531).

  Serial optical sections of tumor cells incubated with diO-LCM for 20 min at RT gave an identical staining pattern as seen in the brain tumor in situ (cf. Fig. 13.1 (bottom right panel)), and showed that LCM became internalized. The process was temperature dependent. When incubated at 37°C, LCM were taken up by C6 tumor cells at the rate of 7 LCM/5 min, while the rate was 4 LCM/5 min at RT. Little uptake by the tumor cells took place at 4°C. These results are illustrated in Fig. 13.5, and are consistent with endocytosis being the main mechanism of LCM uptake by tumor cells (ref. 531).

228

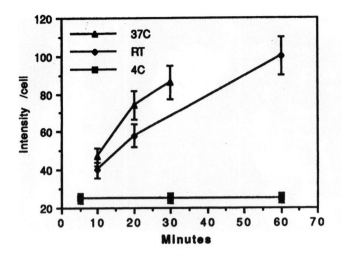

Fig. 13.5. Kinetics of uptake of LCM by tumor cells in culture. C6 cells were incubated with diO-LCM (5 x 10^4 cells/ml) at 3 temperatures for periods ranging from 5 to 60 min. After fixation, the coverslips were examined by fluorescence microscopy. Each experiment was done in triplicate. The data are plotted as the mean values ± S.D. of pooled experiments (n = 3). Intensity is expressed in arbitrary units. (Taken from ref. 531.)

Also consistent with LCM uptake being an active (endocytic) process is the separate finding, in an in vitro kinetic study with both C6 and 9L tumor cells, that both dinitrophenol and sodium azide (i.e., energy blockers) inhibit LCM uptake in both tumor cell lines. Addition of glucose to the medium as an alternate source of energy restored the LCM uptake, indicating an energy-dependent uptake (ref. 534).

13.1.5 LCM are found in acidic compartments in tumor cells in culture: Confocal microscopy using dual-channel recording
In order to further elucidate the fate of LCM in tumor cells, their subcellular localization was determined using chemical markers for various organelles. Permeabilized C6 glioma cells were treated with propidium iodide to label nuclei, WGA for the Golgi apparatus and the plasma membrane, and Rhodamine 123 for mitochondria. None of these subcellular compartments were found to contain LCM. In order to identify intracellular acidic compart-

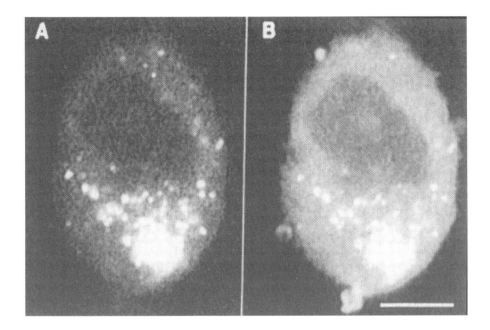

Fig. 13.6. Localization of LCM in intracellular acidic compartments. C6 tumor cells were incubated with diO-LCM and DAMP under conditions described in text. After staining with TR-conjugated anti-DAMP, the cells were examined by confocal microscopy using the dual-channel recording mode for simultaneous detection of diO (A) and TR stains (B). The scale bar represents 10 μm. (Taken from ref. 531.)

ments, C6 cells were incubated with diO-LCM at RT for 20 min, and subsequently with DAMP (Molecular Probes, Eugene, OR) for an additional 30 min. When the C6 tumor cells were examined by confocal fluorescence microscopy using dual-channel recording, it was found that 70% ± 20 of the internalized LCM were associated with acidic compartments (which comprise endosomes and/or lysosomes). An example of this finding is illustrated in Fig. 13.6 where the two fluorochromes, diO label = LCM (Fig. 13.6(A)) and Texas red = acidic compartment (Fig. 13.6(B)), were imaged simultaneously. Eighty percent of internalized LCM had a "mean diameter" of 1.0 μm ± 0.2, indicating that they had clearly aggregated or fused (ref. 531).

13.1.6 Concluding remarks

In all of the animal studies described in Chapters 12 and 13, LCM were always administered to the animals (usually rats) bearing tumor by intravenous injection and, presumably, reach the tumor via the circulation. The interactions of LCM with the perivascular tumor (i.e., 9L gliosarcoma) cells indeed appear to propagate from the circumferential area immediately surrounding blood vessels (Fig. 13.3(A)) to the outer periphery of the tumor (Fig. 13.3(B)) (ref. 531). This pattern reflects the growth properties of the 9L tumor which is known to spread profusely in the brain parenchyma by perivascular invasion (ref. 589). The data reported in this chapter on the distribution of LCM in the brain of rats bearing 9L tumors show that LCM interact preferentially with tumors even when they are remote from the site of implantation. This finding indicates that tissue injury created from implanting tumor cells is not a requisite for LCM attraction (i.e., LCM targeting) (ref. 531). Furthermore, the data reviewed in Chapters 12 and 13 indicate that LCM are rapidly removed from the circulation by tumors. The maximum accumulation of LCM in the tumor area occurs within the first 30 min after intravenous injection (ref. 531).

There is an increased blood flow in brain tumors (ref. 589), and the blood-brain barrier is "leaky" in and around 9L tumors because the blood vessels associated with these (ref. 568), and other (ref. 565-567), tumors are fenestrated. This well-known "leakiness" of tumor capillaries, which in the case of brain tumors includes breaches in the blood-brain barrier (ref. 566,568; cf. Section 12.2), would allow extravasation of small particulate matter (cf. ref. 590-594) or LCM. Once in the tumor area, LCM remain there because of an affinity for tumor cell surface components (cf. ref. 531; see also Chapter 14). At least 4 different types of experimental tumors in rats (C6 glioma, 9L gliosarcoma, Novikoff hepatoma, and Walker-256 carcinosarcoma), as well as several spontaneous tumors in dogs (ref. 570), do interact with LCM in a preferential manner (cf. Chapters 12 and 13), suggesting that LCM affinity may be for tumor cells in general (ref. 531).

Once in contact with the tumor cells, LCM are internalized bringing their content into the cell cytoplasm. Most 9L and C6 tumor cells examined appear to have LCM in their cytoplasm 2 min after LCM i.v. injection. This uptake was observed by confocal laser scanning microscopy both in brain tumors in vivo and in tumor cells in culture. Serial optical sectioning of cells that have been incubated with diO-LCM shows that fluorescent vesicles of the size of LCM are found inside cells, suggesting that LCM remain as single spherical objects during the internalization process, a fact consistent with endocytosis. The time course of internalization and the temperature dependency of the process further support this notion (ref. 531), as does inhibition of LCM uptake by energy blockers, since endocytosis is a temperature-dependent active-uptake process (ref. 448). In rat kidney cells, endocytotic vesicles eventually fuse to generate lysosomes (ref. 595); accordingly, if this were also the case for LCM uptake into tumor cells, one would expect to see the appearance, with time, of large acidic vesicles. The formation of fluorescently-labeled intra-cellular aggregates, and the coincidence of approximately 70% of the internalized LCM with acidic compartments, were both observed. These data indicate that when grown in cell culture, C6 and 9L tumor cells take up LCM via the endocytic pathway (ref. 531).

13.2   EVALUATION OF L.C.M. AS A DELIVERY AGENT OF PACLITAXEL (TAXOL®) FOR TUMOR THERAPY

The experiments reviewed in Section 13.1 demonstrated in detail that LCM can be labeled with lipophilic fluorescent dye and still retain their tumor-targeting properties. Furthermore, these lipophilic-dye-labeled LCM were shown to become internalized by tumor cells both in vitro and in vivo. This tumor-targeting ability of dye-labeled LCM, and their in vivo persistence at the tumor site and/or within the tumor cell for many minutes, suggested a potential use of LCM as a targeted drug-delivery agent or vehicle (ref. 532). Thus, a search was started for lipophilic anticancer drugs (preferably already FDA-approved for clinical use) that could be incorporated into the LCM and carried specifically to the tumor site.

Paclitaxel is a low-molecular-weight, lipophilic compound which is FDA-approved for anticancer therapy. (This compound is referred to in the research literature by both its generic name paclitaxel and the registered tradename Taxol® [Bristol-Myers Squibb Co., N.Y., N.Y.].) Paclitaxel has been shown to exert antitumor activity against many cell lines (ref. 596-599). Paclitaxel induces microtubule assembly in the cytoplasm (ref. 600), blocking mitosis by stabilizing microtubules and not by changing the mass of polymerized microtubules (ref. 601). One issue motivating the study described below is that paclitaxel is toxic to many normal tissues (ref. 597,599), and the traditional oil-based vehicle employed to deliver paclitaxel clinically, cremophor (CRE), is also toxic (ref. 602,603). These drawbacks could potentially be overcome if paclitaxel could be incorporated into LCM and carried to the tumor site, so that the targeting specificity of these LCM could both reduce the systemic effects of paclitaxel and obviate the use of its traditional vehicle, CRE (ref. 532).

13.2.1 Experimental methods

(i) <u>Cell cultures</u>. The 9L gliosarcoma cell line was maintained in minimum essential medium supplemented with 10% fetal calf serum. C6 glioma cells were maintained in F-10 medium supplemented with 10% horse serum and 2.5% fetal calf serum. All tissue culture products were purchased from Life Technologies, Inc. (Bethesda, MD) (ref. 532).

(ii) <u>LCM preparation and labeling</u>. The preparation and labeling of LCM have been described elsewhere (ref. 536). In brief, 20 mg of Filmix® powder (CAV-CON Inc., Farmington, CT) was mixed in 100 ml of isotonic NaCl solution. The resultant saturated solution was then shaken vigorously (by hand, with no sonication necessary) for 10 seconds in air (at room temperature), forming high concentrations of stable gas microbubbles; paclitaxel was added at this step. The suspension was then filtered through a 0.45-μm pore-diameter membrane filter (as done with all Filmix® preparations). The maximum concentration of (bound) paclitaxel was 200 μg/ml for the C6 glioma experiments in rats, and 800 μg/ml for the 9L gliosarcoma experiments also in rats, assuming 100% incorporation of paclitaxel into LCM and assuming no removal of lipids during the filtration step (ref. 532).

(iii) <u>In vitro toxicity of paclitaxel-LCM</u>. C6 cells (10^5 cells) were plated on coverslips in a six-well tissue culture plate. They were ready for use when they reached 50% confluence. Cells at room temperature (six wells each) were treated with paclitaxel-LCM (6 μg/ml), paclitaxel-CRE (6 μg/ml), or LCM alone for 8 or 24 hours. After the cells were rinsed three more times with phosphate-buffered saline, fixed with 4% paraformaldehyde, and washed again, they were ready for immunochemistry (see Section 13.2.2(i)) (ref. 532).

(iv) <u>Tumor models</u>. Two rat-brain tumor models were used: C6 glioma in Sprague-Dawley rats, and 9L gliosarcoma in Fischer 344 rats. The 9L gliosarcoma is syngeneic with the Fischer 344 rats and does not normally become necrotic. This second tumor model has been characterized with respect to the time of death (19.86 ± 1.16 days, n = 14) (ref. 532).

(v) <u>Surgery and tumor implantation</u>. The surgical procedures have been described elsewhere (ref. 525,536). The experiments employed the coordinates from the stereotactic atlas of Pellegrino et al. (ref. 604) and isotonic saline solution containing 2 x 10^4 cultured 9L tumor cells/10 µl or 3 x 10^6 cultured C6 tumor cells/100 µl. The scalp was then sutured closed (ref. 532).

13.2.2 Pharmacological results

(i) <u>Antitumor effect of paclitaxel-LCM in vitro</u>. When cultured C6 glioma cells were treated with paclitaxel-LCM (6 µg/ml), paclitaxel-CRE (6 µg/ml), and LCM alone, paclitaxel-LCM was

TABLE 13.1

Effect of paclitaxel-LCM and paclitaxel-cremophor on cell survival in vitro.[a] (Taken from ref. 532.)

Time of treatment	LCM (cells/field) mean ± S.D.[b]	Paclitaxel-cremophor (cells/field) mean ± S.D.	Paclitaxel-LCM (cells/field) mean ± S.D.
8 hrs	62 ± 18	26 ± 16	19 ± 10
24 hrs	74 ± 18	14 ± 7	13 ± 5

[a] C6 glioma cells were grown in F-10 medium supplemented with 10% horse serum and 2.5% fetal calf serum. Cells were plated on glass coverslips (22 mm^2) and used when they reached 50% confluence. (LCM, lipid-coated microbubbles; S.D., standard deviation.)

[b] Mean of 10 fields counted.

found to be as effective as paclitaxel in cremophor (CRE) in inducing cell death. By 8 hours, 60%-70% of the C6 tumor cells were dead (more with paclitaxel-LCM than paclitaxel-CRE); by 24 hours, approximately 80% were dead with either treatment, whereas LCM alone had no toxic effect on cells (see Table 13.1) (ref. 532).

The effect of paclitaxel on the *morphology* of C6 tumor cells was also examined in culture. C6 cells were treated with LCM alone, paclitaxel-CRE (6 μg/ml), or paclitaxel-LCM (6 μg/ml). Compared to the control (LCM alone), C6 tumor cells treated with either paclitaxel-CRE or paclitaxel-LCM for 24 hours resulted in contracted, rounded tumor cells (ref. 532).

Similar morphological effects are obtained when the above experiment is repeated without cremophor (CRE), but in this case

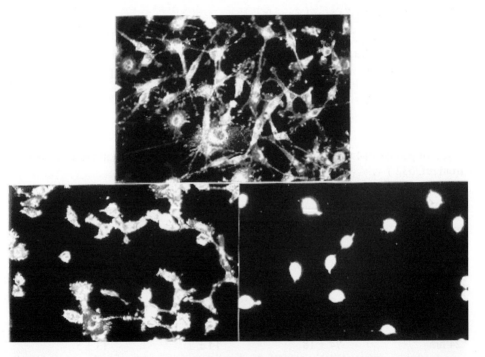

Fig. 13.7. Effects of paclitaxel-loaded LCM on C6 cell morphology. Cells were treated with paclitaxel or paclitaxel-LCM for 8 hours. After treatment the cells were rinsed with saline, fixed, and stained with fluorescent-conjugated WGA to reveal overall cell morphology. Control culture (top panel); paclitaxel (bottom left panel); paclitaxel-LCM (bottom right panel). (Taken from ref. 532.)

the paclitaxel alone has a less pronounced effect than paclitaxel-LCM (at the same paclitaxel concentration) on the C6 cells. These morphological effects are shown in Fig. 13.7. The figure compares the morphology of C6 glioma cells in control culture (Fig. 13.7(A)), paclitaxel-saline and paclitaxel-LCM treated cultures. Overall tumor-cell morphology indicates that paclitaxel-LCM (Fig. 13.7(C)) had a more pronounced effect than paclitaxel-saline (Fig. 13.7(B)) under the conditions tested. In paclitaxel-LCM cultures the C6 cells have retracted processes, while this action is incomplete in paclitaxel-saline cultures after 8 hours. The cell number of paclitaxel-LCM treated culture is also reduced further than that of paclitaxel-saline as compared to control at 8 and 24 hours (cf. ref. 532).

Additional immunochemical staining techniques were utilized in another in vitro experiment in order to further examine the effect of paclitaxel-LCM on the morphology of C6 glioma cells. The cells were grown on glass coverslips and treated with LCM only (Fig. 13.8(C,D)) or paclitaxel-LCM (Fig. 13.8(A,B)) for 8 hours. After treatment, the C6 tumor cells were fixed, and pretreated prior to the immunostaining step (cf. ref. 532). A finer analysis of the ultrastructural abnormality of microtubule structures of C6 cells treated with paclitaxel-LCM (for 8 hrs) was achieved by staining the cells with antibody to tubulin; specifically, the C6 cells were stained both with fluorescein-conjugated wheat germ agglutinin (WGA) to reveal the tumor-cell surface (cf. Section 13.1.2), and with Texas red-conjugated secondary antibody to rabbit anti-tubulin to reveal the presence of microtubules. Cells were examined with confocal microscopy (MRC 600 Confocal System; BioRad, Hercules, CA), where both channels were recorded simultaneously. Fig. 13.8(A,C) show the presence of WGA, while Fig. 13.8(B,D) reveal the presence of tubulin. Shortened microtubules and/or disaligned microtubules were characteristic of the majority of the C6 cells after treatment with paclitaxel-LCM (ref. 532).

(ii) Antitumor effect from single injection, of paclitaxel-LCM, in vivo. Rats bearing C6 brain tumor were treated with paclitaxel in cremophor or paclitaxel in LCM. The treatment consisted of one intravenous (bolus) injection of either 240 μg/kg paclitaxel-CRE or 240 μg/kg paclitaxel-LCM, 4 days after the tumor inoculation. The rats (Sprague-Dawley) were sacrificed 3

236

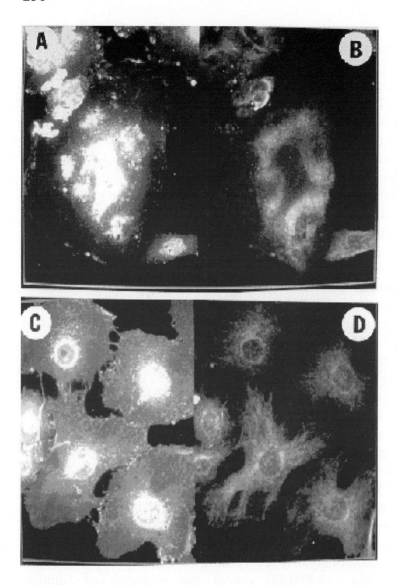

Fig. 13.8. Effect of paclitaxel on the morphology of C6 cells in culture. Cells were treated with LCM only (C,D) or paclitaxel-LCM (A,B) for 8 hours. The cells were stained with fluorescein-conjugated wheat germ agglutinin (WGA) to reveal the cell surface, and with Texas red-conjugated secondary antibody to rabbit anti-tubulin to reveal the presence of microtubules. Both channels were recorded simultaneously. Panels A and C show the presence of WGA, while B and D reveal the presence of tubulin. (Taken from ref. 532.)

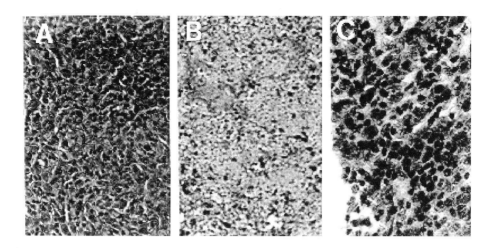

Fig. 13.9. This photomicrograph demonstrates the tumor morphology of Sprague-Dawley rats bearing C6 glioma treated with paclitaxel-CRE (A) and paclitaxel-LCM (B and C). Animals received one intravenous injection of either 240 μg/kg paclitaxel-LCM or 240 μg/kg paclitaxel-CRE, 4 days after the tumor inoculation. Tumors from animals treated with paclitaxel-LCM showed extensive necrosis (B) and lymphocytic infiltration (C). At the same concentration of paclitaxel, the paclitaxel-CRE has no effect on the tumor morphology (A). (Taken from ref. 532.)

days after treatment, and the brain was processed for histology. Fig. 13.9 is a representative view of the C6 tumor from paclitaxel-CRE (Fig. 13.9(A)) and paclitaxel-LCM (Fig. 13.9(B,C)) treated animals, respectively. Paclitaxel given in cremophor has no effect on the tumor morphology; the tumor cells appear healthy (Fig. 13.9(A)). In animals treated with the same dose of paclitaxel but provided in LCM, the situation is dramatically different: Fig. 13.9(B) shows extensive necrosis in the tumor mass; Fig. 13.9(C) shows that the margins of the tumor are literally covered with lymphocytic cells (ref. 532).

     (iii)   <u>Antitumor effect from repeated treatments, of paclitaxel-LCM, in vivo</u>. In the initial series of these experiments, rats (Fischer 344) were inoculated with 9L gliosarcoma cells (2 x 10^4 cells/10 μl). Starting on the 10th day after the tumor inoculation, the experimental animals received one intravenous injection (via tail vein) of paclitaxel-LCM (250 μg/kg) daily for 5 consecu-

238

tive days. The control group received cremophor-LCM. 9L tumors from the rats treated with paclitaxel-LCM displayed extensive cell death (Fig. 13.10(B)) and hemorrhage (Fig. 13.10(C)), whereas tumors from control animals showed no necrosis (Fig. 13.10(A)) (cf. ref. 532).

Determination of the optimal dose for this type of pharmacological study is complex and depends on the tumor type. In the earlier study on Sprague-Dawley rats bearing C6 brain tumors (cf. Section 13.2.2(ii) above), it was found that 240 μg/kg paclitaxel-LCM delivered in one intravenous injection induces morphological changes in the C6 tumor, i.e., necrosis and lymphocytic infiltration, without systemic toxicity or mortality. However, C6 tumor is not proven to be syngeneic with the Sprague-Dawley rat, leaving any conclusions relating to necrosis suspect (ref. 532). In view of that, the treatment experiments were expanded to include the 9L gliosarcoma model, which is syngeneic with Fischer 344 rats. In addition, a regimen of five consecutive daily injections was adopted for the 9L tumor model (see above).

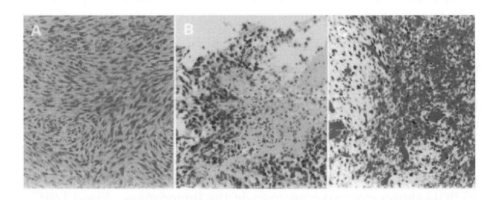

Fig. 13.10 This photomicrograph demonstrates the tumor morphology of Fischer 344 rats bearing 9L gliosarcomas treated with cremophor-LCM (A) and paclitaxel-LCM (B,C). Experimental animals received one i.v. injection of 240 μg/kg paclitaxel-LCM each day for 5 consecutive days starting 10 days after the tumor inoculation. Control animals received cremophor-LCM. Tumors from animals treated with paclitaxel-LCM show extensive cell death (B) and hemorrhage (C). (Taken from ref. 532.)

Earlier data with both Oil Red-O staining and diO-labeled LCM indicated that 9L tumor cells have a much slower LCM uptake than do C6 tumor cells, both in vivo and in vitro (ref. 531). That difference might be a factor in how fast the paclitaxel is consumed and internalized by the tumor cells. A study by Sharma et al. (ref. 605) also suggests that 9L tumor is somewhat paclitaxel-resistant. Accordingly, the daily dose of paclitaxel-LCM was adjusted to 960 µg/kg for subsequent 9L tumor-treatment experiments (ref. 532) (see below).

With this preliminary approximation to a dose/regimen concentration, rats (Fischer 344) employed in the next series of experiments received an intravenous injection daily of paclitaxel-LCM (960 µg/kg) for 5 consecutive days (two and/or three treatment courses, separated by a rest period of 2 days between treatment courses) (see Table 13.2). [All rats weighed 200 to 250 g initially, and were killed when they experienced a progressive weight loss for 3 days or a weight loss of 8 to 10 g per day.] Tukey's Honestly Significant Difference test reveals a significant difference in delayed tumor progression between animals receiving three cycles of treatment versus its respective saline control (P = 0.003) and paclitaxel-CRE control ($P < 0.001$) (ref. 532).

TABLE 13.2

Delayed tumor progression in rats receiving different treatments.[a] (From ref. 532.)

Treatments	"Day of killing"[b]		
	Saline controls (mean ± S.D.)	Paclitaxel-CRE controls (mean ± S.D.)	Paclitaxel-LCM (mean ± S.D.)
Two cycles	21.5 ± 0.70 (n = 2)	19.3 ± 0.57 (n = 3)	24.1 ± 4.24 (n = 4)
Three cycles	20.5 ± 1.73 (n = 5)	19.00 ± 0.89 (n = 6)	27.93 ± 4.17 (n = 15)[c]

[a] S.D., standard deviation; CRE, cremophor; LCM, lipid-coated microbubbles.

[b] "Day of killing", not death, as described in text.

[c] Different from the saline control group and the paclitaxel-CRE group ($P < 0.001$).

(iv) <u>Systemic toxicity of paclitaxel-cremophor</u>. In the 9L tumor model, no deaths from drug toxicity were observed in the treated groups (cf. Table 13.2). Rats receiving LCM alone showed no difference in their growth curve, compared with the control rats. In the animals treated with paclitaxel-CRE, indirect immunofluorescent staining, with anti-PCNA/cyclin, of their gastrointestinal tissues demonstrated a marked decrease of cell proliferation activity when compared with the gastrointestinal tissues obtained from animals treated with paclitaxel-LCM or control animals. This finding is evidence of toxicity to gut epithelium (ref. 532). Moreover, this observation is consistent with the preclinical toxicity evaluation of paclitaxel-CRE by the National Cancer Institute; the major toxic effects are most evident in the tissues with high cell turnover, e.g., hematopoietic, lymphatic, gastrointestinal, and reproductive (male rodents only) (ref. 597). By contrast, the data concerning LCM indicates that the paclitaxel-LCM targeted the tumor without evident influence on other organ systems (ref. 532).

(v) <u>Tissue distribution of LCM in vivo</u>. Ten rats (Fischer 344) bearing 9L gliosarcoma were injected intravenously with LCM ten days after the tumor implantation. Two rats were sacrificed at each of the following time points: 0 to 2 minutes, 1 hour, 3 hours, 6 hours, and 24 hours. Seven organs (brain, liver, kidney, spleen, intestine, muscle, and testes) were removed, quick-frozen in liquid nitrogen, and embedded in O.C.T. compound. Specimens were sectioned with a cryostat at a thickness of 8 μm, stained with Oil Red-O, and examined by light microscopy. Oil Red-O stains the cholesterol esters in the LCM coating (ref. 571). Stained LCM in tissues from the harvested organs, tumor areas in brains (both left side and right side [site of tumor inoculation]), and tissues adjacent to the tumor were each counted using a light microscope equipped with an eyepiece containing a reticle grid (10/10 x 10 units), with a 10X objective (ref. 532).

Table 13.3 indicates the tissue distribution of LCM in rats at different time points after intravenous injection (ref. 532). No microbubbles were detected in: brain within the left hemisphere (i.e., opposite the tumor), spleen, muscle, testes, or intestine at any time after intravenous injection of LCM. Microbubble aggregation in liver increased significantly 3 hours after injection, and by 24 hours, patches of bubbles could be detected. In the kidney, the

TABLE 13.3

Tissue distribution of lipid-coated microbubbles (LCM/mm^2 of tissue) at selected times after injection.[a] (Taken from ref. 532.)

	0-2 min	1 hr	3 hr	6 hr	24.hr
Brain (R)	951	57	0	0	0
Brain (L)	0	0	0	0	0
Liver	1.2	101	525	191	97
Spleen	0	0	0.18	0	0
Testes	0	0	0	0	0
Intestine	0	0	0	0	0
Muscle	0	0	0	0	0
Kidney	1.25	1.25	7.15	11.25	2.75

[a] R, right side; L, left side.

highest concentration of microbubbles was detected between 3 and 6 hours after injection. The microbubbles were concentrated in the renal cortical area (ref. 532). The kidney data are similar to related, preliminary radionuclide data (not shown) obtained by the same research group. It was found that radioactivity from LCM tagged with [^{1}H]cholesterol appeared in urine within 6 hours and showed continued hepatic retention (ref. 532).

(vi) <u>Concluding remarks</u>. LCM differ from other types of coated microbubbles and/or liposomes in many respects; most notably, the LCM coating or surrounding film is a lipid monolayer (comprising saturated glycerides, cholesterol, and cholesterol esters) with <u>no</u> ionic compounds. Because the LCM monolayer coating is uncharged, LCM are unlikely to be captured easily (by the reticuloendothelial system) during circulation; the charge neutrality greatly increases the blood-tissue compatibility (ref.

606). At the same time, this particular lipid-monolayer coating causes (cf. Chapter 14 below) LCM to display the marked tumor-targeting abilities described in Chapters 12 and 13.

Another important property of LCM is that they lodge in and/or adhere to a tumor target, where they remain, without circulation, until they are internalized by the tumor cells and become degraded in the tumor-cell cytoplasm (ref. 531,532; cf. Section 13.1). This finding serves as an added argument for using LCM as a targeted drug-delivery vehicle for various lipophilic anticancer drugs.

Because of the oil solubility of paclitaxel and the thermo-dynamic properties of LCM, paclitaxel is able to attach to LCM. (For the same reasons, a similar ease of incorporation into LCM has been noted for other small, low water-solubility, anticancer drugs [that are otherwise underutilized clinically and/or unpatented] which are being examined currently, such as various oxysterols [like 7β-hydroxycholesterol]; these oxysterols have been reported to inhibit tumor cell proliferation, and gliosis (cf. Section 14.2.2(i)), both in vitro and in vivo (ref. 607-611).)

In summary, both in vitro and in vivo data (reviewed in Section 13.2) indicate that paclitaxel can be lastingly incorporated into LCM. In vivo treatment in rats using two different tumor models (C6 in Sprague-Dawley rats, and 9L in Fischer 344 rats) indicated that intravenously injected paclitaxel-LCM can be delivered to the tumor site, and can exert not only a measurable biological effect but also an antitumor activity (ref. 532).

Chapter 14

PROPOSED MECHANISM OF SELECTIVE L.C.M. UPTAKE BY TUMOR CELLS: ROLE OF LIPOPROTEIN RECEPTOR-MEDIATED ENDOCYTIC PATHWAYS

If the same endocytic LCM-uptake mechanism(s) which has been observed and analyzed in detail with C6 and 9L tumor cells in culture (see Section 13.1) were also operative in vivo, it would indicate that a sizable portion of intravenously injected LCM that bypasses the reticuloendothelial system will then become endocytosed directly by tumor cells (ref. 531). The actual endocytic pathways that are likely to be involved in LCM uptake by tumors are not known, at the present time, due to the lack of any detailed receptor-binding studies with LCM to date. However, a few reasonable candidates for such endocytic pathways emerge upon reviewing parts of an extensive research literature describing significantly enhanced, receptor-mediated endocytosis in many different cancerous cells and solid tumors (see below).

14.1 LOW-DENSITY LIPOPROTEIN (LDL) RECEPTORS, ON TUMOR CELLS, AND L.C.M.

Numerous studies have pointed to an important role for cholesterol during proliferation and progression of cancer (e.g., ref. 612-615). Rapidly dividing cancer cells have two major routes to fulfill their need for cholesterol to form new cell membrane: endogenous synthesis of cholesterol and/or receptor-mediated uptake of exogenous LDL particle-associated cholesterol and cholesterol esters (ref. 612,613,615). Each LDL particle contains a cholesterol ester core surrounded by a polar shell of phospholipids (primarily phosphoglycerides), free cholesterol, and apolipoprotein B (ref. 616-618). Once bound to its cell surface receptor, LDL is internalized by receptor-mediated endocytosis and degraded in lysosomes, and the subsequently released cholesterol may be used for membrane synthesis by the tumor (ref. 619).

The lipid composition of LCM consists of cholesterol esters, free cholesterol, and glycerides (cf. Section 12.1) and is, therefore, similar to the above-mentioned lipid composition of the LDL particle. Based upon this molecular similarity, it appears reasonable to expect that injected LCM would readily bind apolipoprotein B (apo B), either alone or already attached to LDL particles, in the bloodstream. (It is useful to note that LDL particles are formed in large part within peripheral capillaries, i.e., along the capillary endothelium, by the progressive removal of triglycerides from the much larger VLDL particles and by their gradual accumulation of cholesterol esters (ref. 617,620). The well-documented movement of lipid components between physiologic lipoproteins in the blood suggests that a physical interaction between the injected LCM and LDL particles is feasible). The proposed LCM binding of apo B, either alone or already attached to LDL particles, should influence the subsequent biodistribution of those LCM. This expectation derives from two factors: First, apo B is the LDL component which mediates LDL binding to its cell surface receptors (ref. 608), i.e., the LDL receptors, which in turn are involved in receptor-mediated endocytosis (ref. 616,620).

The second factor influencing the expected LCM biodistribution involves the frequently reported finding that many different cancer cells and solid tumors take up LDL more effectively than normal tissues (e.g., ref. 614,615,619,622-625), which many investigators consider is probably a reflection of a higher cholesterol demand of dividing cells as opposed to differentiated cells (e.g., ref. 612-615). (This enhanced receptor-mediated active uptake, or endocytosis, of LDL particles by many cancer cells and tumor types can be due to either increased expression or activity, or both, of the LDL receptors on the cell surface (cf. ref. 613,619,624,626-641).) Some examples of the reported findings are as follows: Leukemic cells isolated from patients with acute myelogenous leukemia have elevated LDL receptor activities compared to normal white blood cells and nucleated bone marrow cells (ref. 626). Gynecologic cancer cells also possess high LDL receptor activity both when assayed in monolayer culture and in membrane preparations from tumor-bearing nude mice (ref. 642). In addition, an enhanced receptor-mediated uptake of LDL by tumor tissue in vivo was demonstrated in animal models and, subsequently, by lung tumors in vivo in

humans (ref. 619). Such enhanced LDL receptor-mediated endocytic uptake of LDL particles, and probably also of LCM (by
analogy [cf. above]) due to their likely bound apo B and/or LDL,
into tumor tissue may therefore explain the extreme rapidity and
high selectivity of the LCM accumulation observed within tumors
(see below).

Specifically, the data reviewed in Chapters 12 and 13 indicate that LCM are rapidly removed from the circulation by the
tumor; the maximum accumulation of LCM in the tumor area
occurs within the first 30 min after administration (ref. 531).
These rapid kinetics for LCM uptake are quite consistent with the
well-documented kinetics long-known for the LDL receptor-
mediated endocytic pathway (ref. 616). For example, Goldstein et
al. (ref. 643) reported that LDL-ferritin bound to coated pits at 4°C
is rapidly internalized when fibroblasts in tissue culture are
warmed to 37°C. In this uptake process, the coated pits invaginate
to form coated endocytic vesicles. After 5 to 10 min at 37°C, LDL-
ferritin is observed in lysosomes as the result of their fusion with
the incoming coated vesicles (ref. 643). The rapid sequence of
events visualized in electron micrographs precisely parallels
biochemically-derived data on the rapid uptake and degradation of
radiolabeled LDL (ref. 644,645).

An additional, related factor that could be contributing to
the rapid removal of LCM from the circulation, by different
tumors, is the fact that vascular endothelial cells participate in LDL
removal from blood (ref. 646). Moreover, it has been reported by
different investigators that growth factors, including tumor-related
growth factors, increase LDL receptor expression on vascular
endothelial cells (e.g., ref. 614,647); this finding raises the
possibility that the endothelial cells arising during cancer-
associated neovascularization may have high numbers of LDL
receptors which could contribute further to the accumulation of
LDL-associated substances, viz. LCM, within tumors. Hence, the
high expression levels of LDL receptors occurring in many types of
tumors (cf. above), and possibly within the tumor capillaries
themselves, could lead to the arrest or assemblage of LCM within
the tumor's vascular supply. By this proposed endocytic mechanism of LCM uptake, the initial interaction between LCM and
lipoprotein receptors would occur intravascularly, a logical
hypothesis for a phenomenon that becomes readily measurable

within 2 minutes following injection of the LCM into the bloodstream (cf. Chapters 12 and 13).

14.2   MULTILIGAND LIPOPROTEIN RECEPTORS

As explained in a review by Krieger and Herz (ref. 648), LDL receptors and most other mammalian cell-surface receptors, which mediate endocytosis, exhibit two common ligand-binding characteristics: high affinity and narrow specificity. The ligand-binding properties of two more recently characterized lipoprotein receptors, i.e., LDL receptor-related protein (LRP) and scavenger receptors, do not conform to a narrow binding specificity. These receptors bind, with high affinity, both lipoprotein and nonlipo-protein ligands and participate in a wide variety of biological processes.

14.2.1 <u>LDL receptor-related protein (LRP), on tumor cells, and LCM</u>

LRP is a member of the "LDL receptor gene family" (ref. 649) and, like the LDL receptor, performs an essential role in the removal of certain lipoprotein particles from the bloodstream. As Heeren et al. (ref. 650) explain, triglycerides are transported mainly by two distinct classes of lipoproteins, the chylomicrons and the very-low-density lipoproteins (VLDL). After assembly in the intestine, chylomicrons are carried via lymph into the bloodstream, where they are transformed at the endothelial surface to remnant lipoproteins through the catalytic action of lipoprotein lipase (for review, see ref. 651,652). After lipolysis, the lipoprotein lipase remains associated with the chylomicron remnants and, in conjunction with apolipoprotein E (apo E) (ref. 653-655), facilitates their clearance by the liver into hepatocytes (ref. 656) via LDL receptors and the LRP (ref. 657-660). (The essential role for both receptors in chylomicron remnant removal in vivo has been demonstrated in gene knockout and gene transfer experiments (ref. 661,662; for review, see ref. 663).)

The remarkable rapidity and specificity of uptake of the chylomicron remnant particles is believed to be highly dependent on their acquisition of apo E in the bloodstream (ref. 664,665); the critical role played by apo E in directing the clearance of chylomicron remnants from the (blood) plasma has been well established by several lines of evidence (ref. 664). Both the LDL receptor (also known as the "apo B,E receptor" (ref. 620)) and the LRP have a high affinity for apo E (ref. 666), and both receptors play an essential role in the receptor-mediated endocytosis of chylomicron remnants (ref. 650,665).

The chylomicron remnant particles themselves, derived from lipolysis of the larger chylomicrons (cf. above), contain the residual triglyceride and all of the cholesterol and cholesterol ester from the original chylomicrons. This lipid composition of the chylomicron remnant particles is similar to the above-described lipid composition of both LDL particles (cf. Section 14.1) and LCM (cf. Section 12.1). Based upon this molecular similarity, it appears reasonable to expect that injected LCM could also readily bind apo E (i.e., as an alternative to apo B) in the bloodstream. In this case, the proposed LCM binding of apo E should influence the subsequent biodistribution of those LCM via two endocytic pathways; specifically, one pathway mediated by the LDL receptor (a.k.a. "apo B,E receptor") (cf. Section 14.1) and the other pathway mediated by the LRP, since both receptor types have a high affinity for apo E (cf. above).

As concerns links between apo E and tumors, Section 14.1 included a review of the increased expression and/or activity of LDL receptors on many cancer cells and tumor types. An analogous, but less widespread, pattern of receptor enhancement has been reported by various investigators for the LRP in tumor cells. One series of reports concerns human glioblastomas (a common type of glioma, or brain tumor) and glioblastoma cell lines: Yamamoto et al. (ref. 667) investigated immunohistochemical localization of LRP, on sequential frozen sections of human glioblastomas, and showed that the neoplastic glial cells and endothelial cells exhibited intense LRP immunoreactivity. Furthermore, they conclude that LRP is overexpressed in glioblastomas, and that LRP may play a role in facilitating glioblastoma invasiveness and neovascularization within tumor tissues (ref. 667). Similarly, Bu et al. reported that human glioblastoma U87 cells express an abundance of LRP, and determined that LRP at the cell surface and along the cellular processes was functional in the binding and endocytosis of its ligands (ref. 668). Finally, in a detailed, recent study by Maletinska et al. (ref. 669), the status of both LRP and LDL receptors was evaluated in seven human glioma (i.e., glioblastoma) cell lines. All were found to exhibit LRP, and cell lines SF-539, U-87 MG, and U-343 MG were particularly rich in this receptor. In addition, the presence of specific saturable LDL receptors was proven in six of the cell lines investigated, which had high concentrations of receptors (128,000-950,000 per cell). These authors conclude this finding suggests that LDL receptors on glioblastoma cells could potentially be useful for targeting antitumor agents; they further conclude that LRP, a multifunctional receptor expressed on glioblastoma cells, also has the possibility for serving as a therapeutic target (ref. 669).

Another major type of tumor cell examined for the presence of LRP was human breast cancer cells. Li et al. (ref. 670) examined six breast cancer cell lines and showed that LRP is expressed at a wide range of levels, i.e., approximately 300 to 6,300 sites per cell. Four of the breast cancer cell lines expressed LRP at over 1,000 sites per cell, and were markedly invasive in their assay (ref. 670).

14.2.2 Scavenger receptors on tumor cells as well as "activated" macrophages: LCM binding, and its relation to certain disease sites

A second category of multiligand lipoprotein receptor, present in varying amounts in different tissues, consists of the scavenger receptors (ref. 671). Various members of this receptor class were initially identified by their capability for rapid, unregulated uptake (by endocytosis) of chemically "modified" LDL by macrophages, leading to massive cholesterol accumulation (ref. 648,671). The similarity of lipid composition between LDL particles and lipid-coated microbubbles (i.e., LCM (cf. Section 12.1)) suggests that the LCM could also resemble a chemically "modified" LDL and, hence, act as a ligand for scavenger receptors. The proposed LCM binding with scavenger receptors would provide a third major endocytic pathway which can influence the biodistribution of LCM (subsequent to their binding of cell-surface receptors).

(i) Tumor cell studies. Indirect support for a proposed third available LCM-uptake mechanism (i.e., scavenger receptor-mediated endocytosis) is found in several studies reporting expression of scavenger receptors on certain types of tumor cells, e.g., hepatoma (ref. 672), renal cell carcinoma (ref. 673), human adrenocortical carcinoma (ref. 631), human breast carcinoma (ref.

615), and histiocytic malignancies (ref. 674-676). Furthermore, when the binding characteristics of these receptors were analyzed in detail in HepG2 (human hepatoma) cells, it was found that in this case the scavenger receptors bound both chemically modified LDL and native LDL (ref. 672). Saturation binding experiments revealed moderate-affinity binding sites for both modified and native LDL, and competition binding studies at 4°C showed that both ligands share common binding site(s). Degradation/association ratios for these ligands show that LDL are very efficiently degraded, and the chemically-modified-LDL degradation/association ratio is equivalent to 60% of the LDL degradation ratio. Both lipoprotein ligands were good cholesterol ester (CE) donors to HepG2 hepatoma cells. Finally, hydrolysis of [^{3}H]CE-lipoproteins in the presence of chloroquine demonstrated that modified and native LDL-CE were mainly hydrolyzed in lysosomes (ref. 672).

(ii) Central-nervous-system injury. While expression of scavenger receptors has been reported for some tumor cell types (cf. Section 14.2.2(i)), there are certain other lesions which far more consistently display increased expression and/or activity of scavenger receptors (see below) on the cell surface. For these particular (noncancerous) disease/injury sites, the above-proposed (cf. Section 14.2.2) third LCM-uptake mechanism (i.e., scavenger receptor-mediated endocytosis) would appear well-suited for targeted drug-delivery therapy, via LCM, of that given disease or injury.

An example of this type of lesion is central-nervous-system (CNS) injury, i.e., brain injury and/or spinal cord injury (see below). As Bell et al. (ref. 677) observed in mouse brain, macrophage scavenger receptor expression on both microglia (the resident macrophages of the CNS) and recruited macrophages was detected 24 hours after an intrahippocampal injection of either lipopolysaccharide or kainic acid (i.e., neurotoxic compounds). Macrophage scavenger receptor expression was also detected in microglia 3 days after optic nerve crush both in the nerve segment distal to the crush site and in the superior colliculus. (The monoclonal antibody 2F8 was used to localize the macrophage scavenger receptor by immunohistochemistry. In control adult mice, microglia did not express the receptor. Also, in the aged mouse brain, the pattern of macrophage scavenger receptor expression was no different from that in the young adult brain.)

Hence, these studies indicate that the increased macrophage scavenger receptor expression, on microglia and recruited macrophages, arising after brain injury could play a role in the clearance of debris during acute neuronal degeneration (ref. 677). At the same time, *this increased receptor expression provides a possible avenue for LCM-targeted drug-delivery treatment of CNS injury sites.*

Furthermore, passage of LCM across the blood-brain barrier would be facilitated by the fact that scavenger receptors (both type I and II, of class A) have been shown, by a reverse transcriptase PCR study of messenger RNA expression, to be present normally in (bovine) brain microvessels (ref. 678). Brain microvessels therefore appear to play an active role in the uptake of native and *modified* LDL (ref. 678); accordingly, these microvessels would be expected to have a similar role in LCM uptake (cf. Sections 12.1 and 14.2.2 above; see also below).

Regarding one specific application of LCM to CNS injury, Kureshi et al. (535) reported the affinity of tail vein-injected LCM to injured rat spinal cord (due to a compressive lesion to the upper thoracic region). The accumulation of LCM in the injured spinal cord was analyzed, after labeling it with a lipid-soluble fluorescent dye (diO), by confocal laser scanning microscopy. It was observed that affinity of LCM for spinal cord injury sites appears to be mediated in the early stages after injury by proliferating macrophages in the necrotic center, and then in later stages by (glial fibrillary acidic protein (GFAP)-positive) activated astrocytes in adjacent white matter. These findings suggested a potential for *using LCM as a delivery vehicle to concentrate lipid-soluble agents in spinal cord injury sites (ref. 535).*

As another application of LCM to CNS injury, Ho et al. (ref. 533) studied the affinity of LCM to the site of a localized (thermal) brain injury. It had been well documented that in response to injury in the CNS, astrocytes are activated which is accompanied by an increased content of GFAP, hypertrophy, and hyperplasia (ref. 679-683). (This process of gliosis (ref. 682) results in scar formation; it has been speculated that the scar may inhibit axonal regeneration (ref. 684)) (ref. 533). Ho et al. observed that the influx of LCM began at the time when GFAP-positive cells began to appear. It seemed likely that LCM are initially attracted to the "reactive" astrocytes, but subsequently the LCM were found to be excluded from the area of scar formation. The LCM distribution detected by fluorescent(diO)-labeled LCM implied that the LCM are rapidly (within 10 min after the intravenous injection of LCM) taken up by some cells other than the astrocytes beginning 7-10 days after the injury. The round shape of these cells suggested that they might be blood borne (i.e., microglia) (ref. 533).

In follow-up to these findings with brain injury, the experiment program with LCM was expanded to next examine the use of LCM to deliver 7β-hydroxycholesterol (7β-OHC) to a radiofrequency (thermal) lesion in the rat brain. With diO-labeled LCM, it was first reaffirmed that LCM target the injury area specifically and not the adjacent normal tissue. With immuno-histochemistry and fluorescent immunochemistry, it was then *demonstrated that the 7β-OHC, delivered in LCM, exerts an antigliosis effect in the rat brain injury model (ref. 534) (see below).*

7β-OHC and other oxysterols have been reported, by other investigators, to inhibit astrogliosis as well as tumor cell prolifera-tion both in vitro and in vivo (ref. 607-611). Several groups have proposed different mechanisms for the cytotoxicity of these compounds (ref. 607,609-611). Kupferberg et al. (ref. 611) first studied the cytotoxic effects of 7β-OHC and its derivatives on reactive astrocyte proliferation. Their group demonstrated that injection of liposomes containing 7β-OHC decreased the choles-terol biosynthesis by 40% in injured rat brain cortex, with a linear relationship between the cholesterogenesis and astroglial proliferation (ref. 607,610,611). They concluded that oxysterols are potent inhibitors of the endogenous cholesterol biosynthesis, and this down-regulation accounts for the inhibition of injury-induced astroglial proliferation, thereby reducing astrocytic reaction (ref. 534).

The data obtained in this follow-up study, based on immunohistochemical staining of GFAP-positive cells, and with fluorescent confocal microscopy techniques, indicate that *both the number of GFAP-expressing astrocytes and the intensity of the staining were reduced when treated with 7β-OHC delivered by the LCM*, while not affected by the same dose of intravenously injected 7β-OHC in saline. Hence, since regenerating axons appear to be physically blocked by scar formation during healing (ref. 684,685), possibly antigliosis agents delivered directly to the injury sites in the optimal concentration (i.e., using targeted drug-delivery via LCM) might assist in promoting functional repair (ref. 534).

The fact that the 7β-OHC injected alone did <u>not</u> exert the same effect indicates that LCM-bound 7β-OHC might be metabolized through a different pathway, i.e., (scavenger) receptor-mediated endocytosis. *As observed previously with the drug paclitaxel (ref. 532; see also Section 13.2), delivery of a given drug via LCM appears to greatly magnify the concentration of that drug reaching the target site. Presumably, the mechanism of this enhanced delivery of 7β-OHC to the injury site by LCM shares common features with the receptor-mediated endocytic pathway mechanisms described earlier for the case of tumor cells (cf. Sections 14.1-14.2.2(i)).* The dissolved free drug, in this case 7β-OHC, would therefore not be expected to reach the desired target sites in as high a concentration compared to when it is carried by the LCM (ref. 534).

(iii) <u>Alzheimer's disease</u>. In Alzheimer's disease the characteristic lesions that develop, called senile plaques, are extracellular deposits principally composed of insoluble aggregates of β-amyloid protein (Aβ) fibrils, infiltrated by reactive microglia and astrocytes (ref. 686). Aβ fibrils exert a cytotoxic effect on neurons, and stimulate microglia to produce neurotoxins, such as reactive oxygen species (ref. 687). Mononuclear phagocytes, including microglia, express scavenger receptors that mediate adhesion and/or endocytosis (ref. 687,688); in particular, microglia have been shown to be intimately associated with amyloid deposits (ref. 688-690) and have also been implicated as scavengers responsible for clearing Aβ fibril deposits of Alzheimer's disease (ref. 688,689). For example, Paresce <u>et al</u>. (ref. 689) reported that microglia in culture rapidly took up fluorescent-labeled Aβ microaggregates into discrete vesicles, which were confirmed to be endosomes; at longer incubation times (30-60 min), Aβ fluorescence became increasingly concentrated in organelles around the nucleus, consistent with delivery to late endosomes and lysosomes. Moreover, the uptake of Aβ microaggregates by the microglia has been shown to be saturable, and is not fluid-phase internalization (ref. 689). Accordingly, El Khoury <u>et al</u>. (ref. 687,691) identified class A scavenger receptors as the main cell-surface receptors mediating the interaction of microglia with β-amyloid fibrils. Adhesion of microglia to β-amyloid fibrils leads to immobilization of these cells on the fibrils, and induces or "activates" them to produce reactive oxygen species. These authors conclude that *microglial scavenger receptors may be novel targets for therapeutic interventions in Alzheimer's disease* (ref. 691). Similarly, a related review by Kalaria (ref. 692) provides additional data supporting the belief that microglia play a major role in the cellular response associated with the pathological lesions of Alzheimer's disease, and concludes that *pharmacological agents which suppress microglial activation may prove a useful strategy to slow the progression of Alzheimer's disease* (ref. 692) (see also ref. 693-696).

In view of the above findings and *since LCM have already been shown (ref. 534; see also Section 14.2.2(ii)) to be an effective in vivo targeted drug-delivery vehicle for the antigliosis drug 7β-hydroxycholesterol (7β-OHC) to experimental injury sites in rat CNS, the LCM/7β-OHC complex may well similarly target the scavenger receptors of Alzheimer's lesions and, thereby, reach the microglial cell surface and suppress microglial activation.* While LCM can be and have been used to carry various lipid-soluble drugs, 7β-OHC is an appealing choice for an antigliosis drug. Specifically, 7β-OHC has a cytotoxic effect on different transformed glial cell lines (and probably also different types of activated, or reactive, glial cells (ref. 534)) which is useful since, as noted earlier (ref. 686,687,691), the senile plaques of Alzheimer's disease are infiltrated by both reactive microglia and astrocytes (ref. 686). This added involvement of astrocytes in Alzheimer's disease is mentioned by various other researchers, such as Malhotra et al. who specifically indicate that the neuritic and amyloid cerebral cortical plaques of Alzheimer's disease are surrounded by reactive astrocytes (ref. 697). Similarly, El Khoury et al. point out that Alzheimer's pathological features include neuronal degeneration, astrogliosis, microgliosis, and the extracellular accumulation of neurotoxic, altered isoforms of β-amyloid (ref. 698) (cf. ref. 693-696).

Based on these considerations, LCM-directed 7β-OHC delivery to the Alzheimer's lesions may inhibit amyloid toxicity via gliosis and, potentially, reduce plaque pathology. Passage of the intravenously injected LCM/7β-OHC complex across the blood-brain barrier to Alzheimer's lesion sites should be readily possible in view of the earlier-mentioned fact that scavenger receptors, both type I and II (of class A), have been shown to be present normally in (bovine) brain microvessels (ref. 678). As stated in Section 14.2.2(ii), brain microvessels therefore appear to play an active role in the uptake of native and *modified* LDL particles (ref. 678); consequently, these microvessels would be expected to have a similar role in uptake of the LCM/7β-OHC complex.

(iv) <u>Atherosclerotic lesions</u>. The macrophage scavenger receptor family of proteins, including class A (type I and II) receptors and class B receptors, recognizes a wide variety of macromolecules (including modified LDL) which, after binding, can be internalized by endocytosis or phagocytosis [or, alternatively, remain at the cell surface and mediate adhesion or lipid transfer through caveolae] (e.g., ref. 699). In pathological states, members of this scavenger-receptor family mediate the recruitment, activation and transformation of macrophages and other cells which appear to be related to the development of not only Alzheimer's disease, but also of atherosclerosis (ref. 699-703).

Lipid accumulation in the blood vessel wall depends on the intracellular uptake by macrophages, which transform into foam cells. Overloaded foam cells finally degenerate, leaving extracellular lipid deposits. The lipid overload of macrophages is brought about by several classes of scavenger receptors that, unlike (native-)LDL receptor, take up *modified* LDL and are not feedback controlled (ref. 701-705). As one example, a major type of class B scavenger receptor is reported to be <u>u</u>pregulated by the pro-atherogenic, modified LDL particles; hence, binding followed by uptake perpetuates a cycle of lipid accumulation and receptor expression (ref. 705). Both class B (ref. 706,707) and class A (ref. 700,701) scavenger receptors are expressed in the lipid-laden macrophages in atherosclerotic lesions (ref. 704,708). Furthermore, Nakagama-Toyama et al. (ref. 708) have reported on the differential distribution of the scavenger receptor types within human coronary atherosclerotic lesions.

In view of the detailed published information available on the presence, functional characteristics, and localization of scavenger receptor populations in atherosclerotic lesions (cf. above) as well as the known structural similarity between modified LDL and LCM (cf. Sections 12.1, 14.2.2-14.2.2(ii)), LCM-directed drug delivery to atherosclerotic lesions may offer a means for targeted drug-delivery therapy of atherosclerosis.

Chapter 15

ENDOCYTOTIC EVENTS VERSUS PARTICLE SIZE: MULTI-
DISCIPLINARY ANALYSES DEMONSTRATE L.C.M. SIZES
ARE MOSTLY SUBMICRON

15.1   CHYLOMICRON REMNANT-LIKE PARTICLE SIZES

In Chapter 14, it was explained that the lipid composition of
LCM (cf. Section 12.1) is similar to the lipid composition of both
chylomicron remnant particles (cf. Section 14.2.1) and LDL particles (cf. Section 14.1). Based upon this molecular similarity, it
was proposed that i.v. injected LCM could readily bind (as do
chylomicron remnants) to apolipoprotein E in the bloodstream (cf.
Section 14.2.1). Both the LDL receptor (a.k.a. "apo B,E receptor") and the LRP (cf. Section 14.2.1) have a high affinity for apo
E (ref. 666), and both receptors play an essential role in the
receptor-mediated endocytosis of chylomicron remnants (ref.
650,665,709). Accordingly, these two endocytic pathways have
been   proposed   (together   with   scavenger   receptor-mediated
endocytosis) to influence LCM distribution in vivo in certain
pathological states (cf. above).

If the remarkable rapidity and specificity of uptake of the
chylomicron remnant particles by these two different endocytic
pathways reflects a similar uptake pattern for LCM as proposed,
then one would probably expect a large portion of the LCM size
distribution to be in the same diameter range as found with
chylomicron remnants. The remnant particles are of particular
interest because their size distribution is known (cf. below) to
display both a wider range (i.e., more polydispersity) and larger
sizes than either LDL or modified LDL particles; accordingly, this
polydispersity provides more information about the size constraints
or limits imposed by the endocytic pathways involved in the active
uptake of chylomicron remnant particles and probably also of
LCM. The diameter range for the native chylomicron remnants has
already been determined by other investigators, from in vivo and in

vitro studies with both reconstituted chylomicron remnant particles and chylomicron remnant-like emulsions, to be approximately 100-200 nm (i.e., ~ 0.1-0.2 µm) (e.g., ref. 710-714). [This reported diameter range for chylomicron remnants also agrees with the fact that the routine clearance of these particles in the body, by the liver (i.e., liver sinusoids → endothelial fenestrae → space of Disse → hepatocytes) (e.g., ref. 715), requires passage through endothelial fenestrae having diameters reported often ranging up to 200 nm (0.2 µm) (ref. 716).] Hence, multidisciplinary analyses of the entire LCM particle (i.e., Filmix® "coated-microbubble/particle" (cf. Section 12.1)) size distribution have been undertaken, with special attention given to the submicron size range, in order to determine what proportion of the LCM population actually falls within this diameter range of 0.1-0.2 µm.

15.2 COMPARISON WITH L.C.M. SIZES: PROPORTION OF L.C.M. POPULATION BETWEEN 0.1-0.2 µm

Collaborative multidisciplinary analyses of the LCM (i.e., Filmix® "coated-microbubble/particle") size distribution, particularly in the submicron size range, were carried out at CT Associates, Inc. (Bloomington, MN) during 2001-2002 (ref. 717). LCM size distributions were measured using three different techniques, i.e., optical particle counting, dynamic light scattering, and scanning electron microscopy; the first of these techniques yielded the most data, some representative examples of which are presented below.

Five different optical particle counters, all of which were manufactured by Particle Measuring Systems (Boulder, CO), were used to measure the size distribution of the LCM. [The five counters utilize different light sources and measure the scattered light over different collection angles. Hence, the raw data collected may vary between these counters, over short data-collection periods and prior to further statistical analysis, since the amount of light scattered is a complicated function of the illuminating light properties, scattering angle, and refractive index. Nonetheless, detailed data analyses confirmed that the five instruments all measured similar concentrations (ref. 717).] Fig. 15.1 displays typical LCM size distributions measured by optical particle counting at various stages during the preparation of the LCM

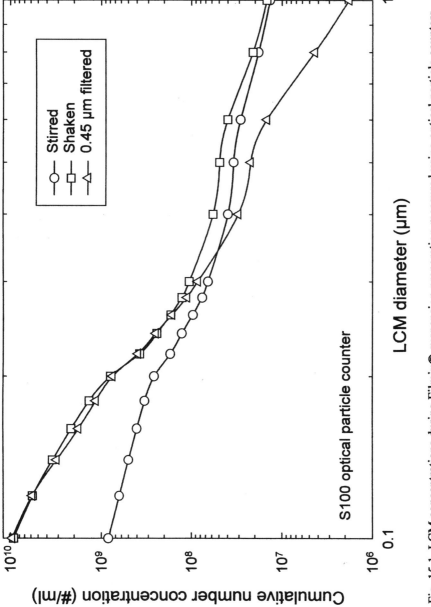

Fig. 15.1. LCM concentrations during Filmix® suspension preparation measured using optical particle counters. (Taken from ref. 717. See text for further discussion.)

258

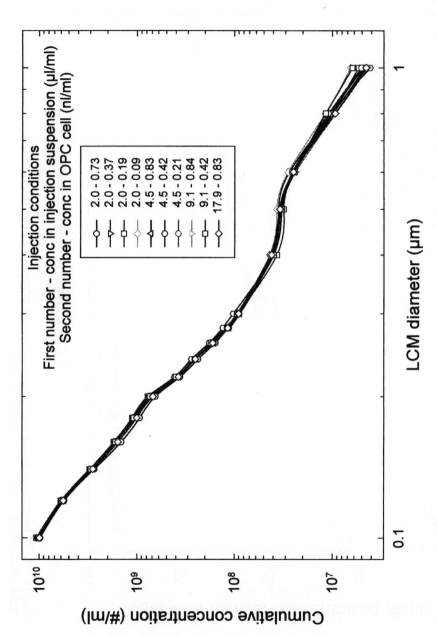

Fig. 15.2. The effect of concentration on LCM particle size distributions. (Taken from ref. 717. See text for further discussion.)

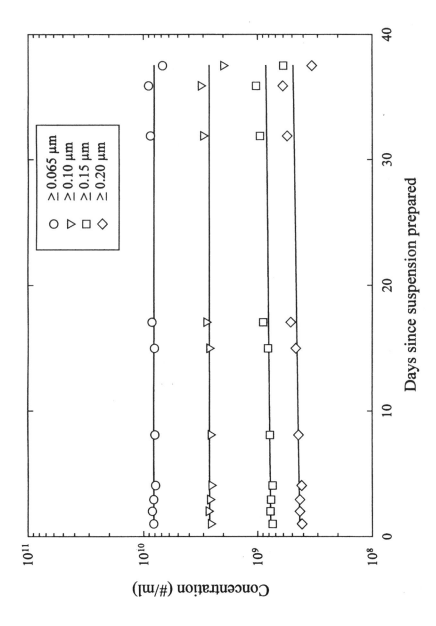

Fig. 15.3. LCM suspension stability over time. (Taken from ref. 717. See text for further discussion.)

suspension. "Cumulative number concentrations" after stirring, shaking, and filtration are shown. The filtered *(Filmix®) suspension contained close to 10^{10} LCM per milliliter ≥ 0.10 μm, with most (~ 90%) of the LCM smaller than 0.2 μm* (ref. 717).

The effect of concentration on the LCM size distributions was also determined using the S100 optical particle counter. It was thought that if the total concentration of LCM material changed, the size distribution might also change as a new equilibrium condition was achieved. The experiment was performed by preparing different dilutions of LCM, then injecting the various dilutions into water at different rates using the dilution/flow system set up for the S100 instrument. In this way, multiple concentrations could be obtained in both the injecting and final suspensions. Fig. 15.2 compares the size distribution of the original Filmix® agent measured under the different sets of dilution conditions. All of the measurements were found to be essentially identical, indicating that the LCM did <u>not</u> change size when subjected to the different concentration conditions (ref. 717).

In addition, the effect of Filmix® suspension age on the size distribution was determined by measuring the LCM size distribution at different times over 37 days, using the M65 optical particle counter. The results are displayed in Fig. 15.3. <u>No</u> change in the LCM size distribution with time (over at least one month) was evident (ref. 717).

In summary, the size distribution of LCM in deionized water was measured using several analytical techniques. The Filmix® agent was found to contain close to 10^{10} LCM/ml ≥ 0.1 μm, when measured using optical particle counters. A large majority (~ 90%) of the LCM (i.e., Filmix® "coated-microbubbles/particles") were smaller than 0.2 μm in diameter.

REFERENCES

1. R. Dean, The formation of bubbles, J. Appl. Phys. 15 (1944) 446-451.
2. E.N. Harvey, D.K. Barnes, W.D. McElroy, A.H. Whiteley, D.C. Pease and K.W. Cooper, Bubble formation in animals, 1, Physical factors, J. Cell. Comp. Physiol. 24 (1944) 1-22.
3. D.C. Pease and L.R. Blinks, Cavitation from solid surfaces in the absence of gas nuclei, J. Phys. Colloid Chem. 51 (1947) 556-567.
4. F.E. Fox and K.F. Herzfeld, Gas bubbles with organic skin as cavitation nuclei, J. Acoust. Soc. Am. 26 (1954) 984-989.
5. K.S. Iyengar and E.G. Richardson, Measurements on the air nuclei in natural water which give rise to cavitation, Brit. J. Appl. Phys. 9 (1958) 154-158.
6. R.T. Knapp, Cavitation and nuclei, Trans. Am. Soc. Mech. Engin. 80 (1958) 1315-1324.
7. M. Strasberg, Onset of ultrasonic cavitation in tap water, J. Acoust. Soc. Am. 31 (1959) 163-176.
8. D.M.J.P. Manley, Change of size of air bubbles in water containing a small dissolved air content, Brit. J. Appl. Phys. 11 (1960) 38-42.
9. W.R. Turner, Microbubble persistence in fresh water, J. Acoust. Soc. Am. 33 (1961) 1223-1232.
10. N.H. Langton and P. Vaughan, Cavitation and the ultrasonic degradation of high polymers, Brit. J. Appl. Phys. 13 (1962) 478-482.
11. M.M. Hasan and K.S. Iyengar, Size and growth of cavitation bubble nuclei, Nature 199 (1963) 995-996.
12. D. Messino, D. Sette and F. Wanderlingh, Effects of uranium salts on sound cavitation in water, J. Acoust. Soc. Am. 35 (1963) 926-927.
13. D. Messino, D. Sette and F. Wanderlingh, Effects of solid impurities on cavitation nuclei in water, J. Acoust. Soc. Am. 41

(1967) 573-583.

14.   E. Webster, Cavitation, Ultrasonics 1 (1963) 39-48.

15.   L.H. Bernd, Study of the surface films of gas nuclei, ONR Contract No. NONR-4296(00), Part I: AD608094, G.E. - ATL, 1964.

16.   L.H. Bernd, Cavitation, tensile strength, and the surface films of gas nuclei, in R.D. Cooper and S.W. Doroff (Eds.), Sixth Symposium on Naval Hydrodynamics, Office of Naval Research, Wash. D.C., 1966, pp. 77-114.

17.   L.H. Bernd, Study of the surface films of gas nuclei, Part IV, ONR Contract No. NONR-4296(00), Part IV: AD829881, G.E. - RDC, 1967, 50 pp.

18.   H.G. Flynn, Physics of acoustic cavitation in liquids, in W.P. Mason (Ed.), Physical Acoustics: Principles and Methods, Vol. I, Part B, Academic Press, New York, 1964, pp. 57-172.

19.   D. Sette and F. Wanderlingh, Thermodynamic theory of bubble nucleation induced in liquids by high-energy particles, J. Acoust. Soc. Am. 41 (1967) 1074-1082.

20.   F.G. Hammitt, D.M. Ericson, J.F. Lafferty and M.J. Robinson, Gas content, size, temperature, and velocity effects on cavitation inception in a venturi, Amer. Soc. Mech. Engin. paper 67-WA/FE-22, 1968.

21.   C.L. Darner, Sonic cavitation in water, Naval Research Laboratory Underwater Sound Research Division Report 7131, ADA031182, NRL-USRD, 1970.

22.   L.R. Gavrilov, On the size distribution of gas bubbles in water, Sov. Phys. Acoust. 15 (1969) 22-24.

23.   A.T.J. Hayward, The role of stabilized gas nuclei in hydrodynamic cavitation inception, J. Phys. D 3 (1970) 574-579.

24.   J.W. Holl, Nuclei and cavitation, J. Basic Engng. 92 (1970) 681-688.

25.   M.G. Sirotyuk, Stabilization of gas bubbles in water, Sov. Phys. Acoust. 16 (1970) 237-240.

26.   A. Keller, The influence of the cavitation nucleus spectrum on cavitation inception, investigated with a scattered light counting method, J. Basic Engng. 94 (1972) 917-925.

27.   A.J. Acosta and B.R. Parkin, Cavitation inception - a selective review, J. Ship Res. 19 (1975) 193-205.

28.   H. Kimoto, Cavitation in purified water, in R.L. Waid (Ed.), Cavitation and Polyphase Flow Forum, American Society of Mechanical Engineers, New York, 1977, pp.1-2.

29. D.E. Yount, T.D. Kunkle, J.S. D'Arrigo, F.W. Ingle, C. Yeung and E.L. Beckman, Stabilization of gas cavitation nuclei by surface-active compounds, Aviat. Space Environ. Med. 48 (1977) 185-191.

30. W.L. Ramsey, Dissolved oxygen in shallow near-shore water and its relation to possible bubble formation, Limnol. Oceanogr. 7 (1962) 453-461.

31. H. Medwin, In situ acoustic measurements of bubble populations in coastal ocean waters, J. Geophys. Res. 75 (1970) 599-611.

32. H. Medwin, In situ acoustic measurements of microbubbles at sea, J. Geophys. Res. 82 (1977) 971-976.

33. B.D. Johnson and R.C. Cooke, Bubble populations and spectra in coastal waters: a photographic approach, J. Geophys. Res. 84 (1979) 3761-3766.

34. D.M. LeMessurier, Supersaturation and "preformed nuclei" in the etiology of decompression sickness, paper presented at the Second International Meeting on Aerospace Medicine, Melbourne, 1972 (unpublished).

35. R.H. Strauss, Bubble formation in gelatin: implications for prevention of decompression sickness, Undersea Biomed. Res. 1 (1974) 169-174.

36. E.A. Cassell, K.M. Kaufman and E. Matijevic, The effects of bubble size on microflotation, Water Res. 9 (1975) 1017-1024.

37. J.S. D'Arrigo, Glycoprotein surfactants stabilize long-lived gas microbubbles in the environment, in J.E. Zajic (Ed.), Microbial-Enhanced Oil Recovery, PennWell Press, Tulsa, 1983, pp. 124-140.

38. W.D. Garrett, Stabilization of air bubbles at the air-sea interface by surface-active material, Deep- Sea Res. 14 (1967) 661-672.

39. N.L. Jarvis, W.D. Garrett, M.A. Scheiman and C.O. Timmons, Surface chemical characterization of surface-active material in seawater, Limnol. Oceanogr. 12 (1967) 88-96.

40. N.L. Jarvis, Adsorption of surface-active material at the sea-air interface, Limnol. Oceanogr. 12 (1967) 213-221.

41. B.D. Johnson and R.C. Cooke, Organic particle and aggregate formation resulting from the dissolution of bubbles in seawater, Limnol. Oceanogr. 25 (1980) 653-661.

42. B.D. Johnson and R.C. Cooke, Generation of stabilized micro-bubbles in seawater, Science 213 (1981) 209-211.

43. D.C. Blanchard, Sea-to-air transport of surface active material,

Science 149 (1964) 396-397.

44.　W.R. Barger and W.D. Garrett, Surface active organic material in the marine atmosphere, J. Geophys. Res. 75 (1970) 4561-4566.

45.　D.C. Blanchard, Jet drop enrichment of bacteria, virus, and dissolved organic material, Pure Appl. Geophys. 116 (1978) 302-308.

46.　A. Detwiler and D.C. Blanchard, Aging and burating bubbles in trace-contaminated water, Chem. Engng. Sci. 33 (1978) 9-13.

47.　A. Detwiler, Surface-active contamination on air bubbles in water, in K.L. Mittal (Ed.), Surface Contamination: Genesis, Detection and Control, Vol. 2, Plenum, New York, 1979, pp. 993-1007.

48.　S.R. Burger and D.C. Blanchard, The persistence of air bubbles at a seawater surface, J. Geophys. Res. 88 (1983) 7724-7726.

49.　J.S. D'Arrigo, Dissolved carbon dioxide: evidence for a significant role in the etiology of decompression sickness, in J.W. Miller (Ed.), Proceedings of the Fourth Joint Meeting of the Panel on Diving Physiology and Technology, U.S.-Japan Cooperative Program in Natural Resources (UJNR), Buffalo, N.Y., 1977, pp. 88-100 (limited distribution).

50.　P.J. Mulhearn, Distribution of microbubbles in coastal waters, J. Geophys. Res. 86 (1981) 6429-6434.

51.　R.T. Roberts, Glycoproteins and beer foam, Eur. Brewing Conv. Proc. 15 (1975) 453-464.

52.　J.S. Pierce, The role of positive and negative factors in head retention, Inst. Brewing, Aust. & N.Z. Sect., Proc. Conv. 15 (1978) 51-65.

53.　A.L. Whitear, Basic factors that determine foam stability, Inst. Brewing, Aust. & N.Z. Sect., Proc. Conv. 15 (1978) 67-75.

54.　J.S. D'Arrigo, in Hearing before the Subcommittee on Aerospace Technology and National Needs of the Committee on Aeronautical and Space Sciences, United States Senate, U.S. Government Printing Office, Wash. DC, 1976, p. 287.

55.　Y. Mano and J.S. D'Arrigo, The relationship between CO_2 levels and decompression sickness: implications for disease prevention, Aviat. Space Environ. Med. 49 (1978) 349-355.

56.　J.S. D'Arrigo and Y. Mano, Bubble production in agarose gels subjected to different decompression schedules, Undersea Biomed. Res. 6 (1979) 93-98.

57.　D.M. Oldenziel, A new instrument in cavitation research: the cavitation susceptibility meter, Trans. Am. Soc. Mech. Engin. 104 (1982) 136-142.

58.   D.M. Oldenziel. New instruments in cavitation research, in W.B. Morgan and B.R. Parkin (Eds.), International Symposium on Cavitation Inception, American Society of Mechanical Engineers, New York, 1979, pp. 111-124.

59.   M.L. Billet and E.M. Gates, A comparison of two optical techniques for measuring cavitation nuclei, Trans. Am. Soc. Mech. Engin. 103 (1981) 8-13.

60.   E.A. Weitendorf, Conclusions from full scale and model investigations of the free air content and of the propeller excited hull pressure amplitudes due to cavitation, in W.B. Morgan and B.R. Parkin (Eds.), International Symposium on Cavitation Inception, American Society of Mechanical Engineers, New York, 1979, pp. 207-218.

61.   A.J. Rubin and E.A. Cassell, Microflotation of bacteria, Proc. 14th Southern Water Resources Pollut. Control Conf., Univ. of North Carolina, 1965.

62.   A.J. Rubin, E.A. Cassell, O. Henderson, J.D. Johnson and J.C. Lamb, Microflotation: new low gas flow rate foam separation technique for bacteria and algae, Biotech: Bioengng. 8 (1966) 135-150.

63.   A.J. Rubin and S.C. Lackey, Effect of coagulation on the microflotation of *B. Cereus*, J. Am. Water Works Assoc. 60 (1971) 1156-1166.

64.   E.A. Cassell, A.J. Rubin, H.B. LaFever and E. Matijevic, Removal of organic colloids by microflotation, Proc. 23rd Ind. Waste Conf., Purdue Univ., 1968.

65.   T.D. Buzzell, The chemical and physical aspects of humic acid removal by microflotation and settling, Masters Thesis,Clarkson College of Technology, 1969.

66.   F.J. Mangravite, E.A. Cassell and E. Matijevic, The microflotation of silica, J. Colloid Interface Sci. 39 (1972) 357-366.

67.   A.J. Rubin and S.F. Erickson, Effect of coagulation and restabilization on the microflotation of illite, Water Res. 5 (1971) 437-444.

68.   A.J. Rubin and D.C. Haberkost, Coagulation and flotation of colloidal titanium dioxide, Sep. Sci. 8 (1973) 363-373.

69.   S.B. Blabac, The removal of polystyrene latex from water by

266

microflotation, Masters Thesis, Clarkson College of Technology, 1970.

70. E.A. Cassell, E. Matijevic, F.J. Mangravite, T.D. Buzzell and S.B. Blabac, Removal of colloidal pollutants by microflotation, A.I.Ch.E. Journ. 17 (1971) 1486-1492.

71. D.G. DeVivo and B.L. Karger, Studies in the flotation of colloidal particulates: effects of aggregation in the flotation process, Sep. Sci. 5 (1970) 145-167.

72. J. Kanwisher, On the exchange of gases between the atmosphere and the sea, Deep-Sea Res. 10 (1963) 195-207.

73. W.H. Sutcliffe, Jr., E.R. Baylor and D.W. Menzel, Sea-surface chemistry and Langmuir circulation, Deep-sea Res. 10 (1963) 233-243.

74. B.D. Johnson, Nonliving organic particle formation from bubble dissolution, Limnol. Oceanogr. 21 (1976) 444-446.

75. D.C. Blanchard and A.H. Woodcock, Bubble formation and modification in the sea and its meteorological significance, Tellus 9 (1957) 145-158.

76. D.C. Blanchard and L.D. Syzdek, Concentration of bacteria in jet drops from burating bubbles, J. Geophys. Res. 77 (1972) 5087-5099.

77. G.T. Wallace and R.A. Duce, Concentration of particulate trace metals and particulate organic carbon in marine surface waters by a bubble flotation mechanism, Mar. Chem. 3 (1975) 157-181.

78. E.R. Baylor, M.B. Baylor, D.C. Blanchard, L.D. Syzdek and C. Appel, Virus transfer from surf to wind, Science 198 (1977) 575-580.

79. G.A. Riley, Organic aggregates in sea water and the dynamics of their formation and utilization, Limnol. Oceanogr. 8 (1963) 372-381.

80. P.J. Wangersky, Particulate organic carbon: sampling variability, Limnol. Oceanogr. 19 (1974) 980-984.

81. J.R. Wheeler, Formation and collapse of surface films, Limnol. Oceanogr. 20 (1975) 338-342.

82. R.W. Sheldon, T.P. Evelyn and T.R. Parsons, On the occurrence and formation of small particles in seawater, Limnol. Oceanogr. 12 (1967) 367-375.

83. E. Batoosingh, G.A. Riley and B. Keshwar, An analysis of experimental methods for producing particulate organic matter in sea water by bubbling, Deep-Sea Res. 16 (1969) 213-219.

84. L. Liebermann, Air bubbles in water, J. Appl. Phys. 28 (1957) 205-211.

85. D.C. Blanchard, Bubble scavenging and the water-to-air transfer of organic material in the sea, Adv. Chem. Ser. 145 (1975) 360-387.

86. W.D. Garrett, The organic chemical composition of the ocean surface, Deep-Sea Res. 14 (1967) 221-227.

87. W.D. Garrett, Collection of slick-forming material from the sea surface, Limnol. Oceanogr. 10 (1965) 602-605.

88. G.W. Harvey, Microlayer collection from the sea surface: a new method and initial results, Limnol. Oceanogr. 11 (1966) 608-613.

89. P.M. Williams, Organic acids in Pacific Ocean waters, Nature 189 (1961) 219-220.

90. T.R. Parsons, Suspended organic matter in sea water, Progr. Oceanogr. 1 (1963) 203-239.

91. V.P. Glotov, P.A. Kolobaev and G.G. Neuimin, Investigation of scattering of sound by bubbles generated by an artificial wind in sea water and the statistical distribution of bubble sizes, Sov. Phys. Acoust. 7 (1962) 341-345.

92. E.J. Hoffman and R.A. Duce, The organic carbon content of marine aerosols collected on Bermuda, J. Geophys. Res. 79 (1974) 4474-4477.

93. A.T. Wilson, Surface of the ocean as a source of air-borne nitrogenous material and other plant nutrients, Nature 184 (1959) 99-101.

94. E. Jones, Contribution of rainwater to the nutrient economy of soil in northern Nigeria, Nature 188 (1960) 432.

95. R.B. Miller, The chemical composition of rainwater at Taita, New Zealand, 1956-1958, New Zealand J. Sci. 4 (1961) 844-853.

96. D.C. Blanchard and L.D. Syzdek, Importance of bubble scavenging in the water-to-air transfer of organic material and bacteria, J. Recherches. Atmos. 8 (1974) 529-540.

97. A.F. Carlucci and H.F. Bezdek, On the effectiveness of a bubble for scavenging bacteria from sea water, J. Geophys. Res. 77 (1972) 6608-6610.

98. J.A. Quinn, R.A. Steinbrook and J.L. Anderson, Breaking bubbles and the water-to-air transport of particulate matter, Chem. Engng. Sci. 30 (1975) 1177-1184.

99. F. MacIntyre, Flow patterns in breaking bubbles, J. Geophys. Res. 77 (1972) 5211-5228.

268

100. E. Hoffman, Ph.D. Thesis, University of Rhode Island, 1975.

101. G.T. Wallace, Jr., Ph.D. Thesis, University of Rhode Island, 1976.

102. R. Lemlich (Ed.), Adsorptive Bubble Separation Techniques, Academic Press, New York, 1972.

103. B.V. Derjaguin, V.A. Fedoseyev and L.A. Rosenzweig, Investigation of the adsorption of cetyl alcohol vapor and the effect of this phenomenon on the evaporation of water drops, J. Colloid Interface Sci. 22 (1966) 45-50.

104. J.E. Jiusto, Investigation of warm fog properties and fog modification concepts, Cornell Aeronautical Lab. Rep. Rm-17788-P-4, Buffalo, N.Y., 1964.

105. H.L. Rosano and V.K. La Mer, The rate of evaporation of water through monolayers of esters, acids and alcohols, J. Phys. Chem. 60 (1956) 348-353.

106. G.T. Barnes and V.K. La Mer, The evaporation resistances of monolayers of long-chain acids and alcohols and their mixtures, in V.K. La Mer (Ed.), Retardation of Evaporation by Monolayers: Transport Processes, Academic Press, New York, 1962, pp. 9-33.

107. R.R. Cruse, Structural geometry in the selection of retardants and dispersants for use in water evaporation suppression, in V.K. La Mer (Ed.), Retardation of Evaporation by Monolayers: Transport Processes, Academic Press, New York, 1962, pp. 219-233.

108. A. Goetz, The constitution of aerocolloidal particulates above the ocean surface, Proc. Int. Conf. Cloud Phys., Tokyo and Sapporo, Japan, 1965, pp. 42-45.

109. D.C. Blanchard, The electrification of the atmosphere by particles from bubbles in the sea, Progr. Oceanogr. 1 (1963) 71-202.

110. W.C. Jacobs, A preliminary report on a study of atmospheric chlorides, Mon. Weather Rev. 65 (1937) 147-151.

111. A.H. Woodcock, C. Kientzler, A.B. Arons and D.C. Blanchard, Giant condensation nuclei from bursting bubbles, Nature 172 (1953) 1144.

112. J.A. Day, Production of droplets and salt nuclei by the bursting of air bubble films, Q. J. R. Meteorol. Soc. 90 (1964) 72-78.

113. R.J. Cipriano and D.C. Blanchard, Bubble and aerosol spectra produced by a laboratory "breaking wave", J. Geophys. Res. 86 (1981) 8085-8092.

114. T.D. Kunkle, Bubble nucleation in supersaturated fluids, Ph.D. Thesis, University of Hawaii, 1979, 108 pp.

115. M. Blank and F.J.W. Roughton, Permeability of monolayers to carbon dioxide, Trans. Faraday Soc. 56 (1960) 1832-1841.

116. M. Blank, The permeability of monolayers to several gases, in V.K. La Mer (Ed.), Retardation of Evaporation by Monolayers: Transport Processes, Academic Press, New York, 1962, pp. 75-95.

117. J.G. Hawke and A.E. Alexander, The influence of surface-active compounds upon the diffusion of gases across the air-water interface, in V.K. La Mer (Ed.), Retardation of Evaporation by Monolayers: Transport Processes, Academic Press, New York, 1962, pp. 67-73.

118. J.G. Hawke and A.G. Parts, A coefficient to characterize gaseous diffusion through monolayers at the air/water interface, J. Colloid Sci. 19 (1964) 448-456.

119. F. Goodridge and I.D. Robb, Mechanism of interfacial resistance in gas absorption, Ind. Engng. Chem. Fundls. 4 (1965) 49-55.

120. H.M. Princen, J.T.G. Overbeek and S.G. Mason, The permeability of soap films to gases, II, A simple mechanism of monolayer permeability, J. Colloid Interface Sci. 24 (1967) 125-130.

121. M. Linton and K.L. Sutherland, Solution of oxygen through a monolayer, Australian J. Appl. Sci. 9 (1958) 18-23.

122. N.P. Berezina and N.V. Nikolaeva-Fedorovich, On the purity of water in electrochemical studies, Elektrokhimiya 3 (1967) 3-7.

123. V.G. Levich, Physicochemical Hydrodynamics, Fizmatgiz, Moscow, 1959.

124. M.G. Sirotyuk, Experimental investigation of ultrasonic cavitation, in L.D. Rozenberg (Ed.), Physics and Technology of High-Intensity Ultrasound, Vol. 2, High-Intensity Ultrasonic Fields, Nauka, Moscow, 1968.

125. D.M. Oldenziel, A set of instruments useful for liquid quality control during cavitation research, J. Fluids Engng. 104 (1982) 443-450.

126. R.A. Robinson and R.H. Stokes, Electrolyte Solutions: The Measurement and Interpretation of Conductance, Chemical Potential and Diffusion in Solutions of Simple Electrolytes, 2nd ed., Butterworths, London, 1959, 571 pp.

127. F. Avellan and F. Resch, A scattering light probe for the

270

measurement of oceanic air bubble sizes, Int. J. Multiphase Flow 9 (1983) 649-663.

128. S.H. Maron and J.B. Lando, Fundamentals of Physical Chemistry, Macmillan, New York, 1974, 853 pp.

129. H.B. Bull, An Introduction to Physical Biochemistry, 2nd ed., F.A. Davis, Philadelphia, 1971, 469 pp.

130. P.S. Epstein and M.S. Plesset, On the stability of gas bubbles in liquid-gas solutions, J. Chem. Phys. 18 (1950) 1505-1509.

131. H. Medwin, Acoustic fluctuations due to microbubbles in the near-surface ocean, J. Acoust. Soc. Amer. 56 (1974) 1100-1104.

132. A.P. Keller and E.A. Weitendorf, A determination of the free air content and velocity in front of the "Sydney-Express" propeller in connection with pressure fluctuation measurements, in Twelfth Symposium on Naval Hydrodynamics: Boundary Layer Stability and Transition, Ship Boundary Layers and Propeller-Hull Interaction, Cavitation, and Geophysical Fluid Dynamics, National Academy of Sciences, Wash. DC, 1979, pp. 300-318.

133. D.C. Gordon, Jr., A microscopic study of organic particles in the North Atlantic Ocean, Deep-Sea Res. 17 (1970) 175-185.

134. K.L. Carder, G.F. Beardsley and H. Pak, Particle size distributions in the Eastern Equatorial Pacific, J. Geophys. Res. 76 (1971) 5070-5077.

135. D.E. Yount and R.H. Strauss, Bubble formation in gelatin: a model for decompression sickness, J. Appl. Phys. 47 (1976) 5081-5089.

136. D.H. LeMessurier, T.N. Smith and W.R. Wood, Diffusion and nucleation of gas in gel and some implications for the development of decompression, Undersea Biomed. Res. 6 (1979) 175-188.

137. T.N. Smith, D.H. LeMessurier and W.R. Wood, paper from the Dept. of Chem. Engng., University of Adelaide, 1980 (limited distribution).

138. J.S. D'Arrigo, An improved method for studying the physical chemistry of bubble formation, Biophys. J. 17 (Feb. 1977) 302a.

139. J.S. D'Arrigo, Improved method for studying the surface chemistry of bubble formation, Aviat. Space Environ. Med. 49 (1978) 358-361.

140. A. Veis, The Macromolecular Chemistry of Gelatin, Academic Press, New York, 1964.

141. S. Hjerten, Agarose as an anticonvection agent in zone electro-

phoresis, Biochem. Biophys. Acta 53 (1961) 514-517.

142.   C. Araki, Structure of the agarose constituent of Agar-agar, Bull. Chem. Soc. Japan 29 (1956) 543-544.

143.   M. Duckworth and W. Yaphe, Preparation of agarose by fractionation from the spectrum of polysaccharides in agar, Anal. Biochem. 44 (1971) 636-641.

144.   M. Duckworth and W. Yaphe, The structure of agar, Part I, Fractionation of a complex mixture of polysaccharides, Carbohyd. Res. 16 (1971) 189-197.

145.   K.B. Guiseley and D.W. Renn, Agarose: Purification, Properties, and Biomedical Applications, FMC Corp., Rockland, 1975, 34 pp.

146.   N.S. Radin, L. Hof and C. Seidl, Lipid contaminants: polypropylene apparatus and vacuum pumps, Lipids 3 (1968) 192.

147.   J.S. D'Arrigo, Evidence that nonionic surfactants surround gas cavitation nuclei, in Proceedings of the 1979 Aerospace Medical Association Meeting, 1979 (unpublished; limited distribution).

148.   J.S. D'Arrigo, Axonal surface charges: evidence for phosphate structure, J. Membrane Biol. 22 (1975) 255-263.

149.   J.T. Davies and E.K. Rideal, Interfacial Phenomena, Academic Press, New York, 1963, 480 pp.

150.   W.C. Griffin, Classification of surface active agents by "HLB", J. Soc. Cosmetic Chem. 1 (1949) 311-326.

151.   H. Hauser, M.C. Phillips and M.D. Barratt, Differences in the interaction of inorganic and organic (hydrophobic) cations with phosphatidylserine membranes, Biochem. Biophys. Acta 413 (1975) 341-353.

152.   L.G. Sillen and A.E. Martell, Stability Constants of Metal-Ion Complexes, Supplement 1, Special publication no. 25, The Chemical Society, London, 1971, 865 pp.

153.   J.S. D'Arrigo, Physical properties of the nonionic surfactants surrounding gas cavitation nuclei, J. Chem. Phys. 71 (1979) 1809-1813.

154.   T.M. Doscher, G.E. Myers and D.C. Atkins, Jr., The behavior of nonionic surface active agents in salt solutions, J. Colloid Sci. 6 (1951) 223-235.

155.   F.A. Long and W.F. McDevit, Activity coefficients of nonelectrolyte solutes in aqueous salt solutions, Chem. Rev. 51 (1952) 119-169.

156.   C.F. Jelinek and R.L. Mayhew, Nonionic surfactants: their

chemistry and textile uses, Textile Res. J. 24 (1954) 765-778.

157. W.N. Maclay, Factors affecting the solubility of nonionic emulsifiers, J. Colloid Sci. 11 (1956) 272-285.

158. F.E. Bailey, Jr. and R.W. Callard, Some properties of poly(ethylene oxide) in aqueous solution, J. Appl. Polym. Sci. 1 (1959) 56-62.

159. K. Durham, Properties of detergent solutions - amphipathy and absorption, in K. Durham (Ed.), Surface Activity and Detergency, Macmillan, London, 1961, pp. 1-28.

160. P. Becher, Nonionic surface-active compounds, V, Effect of electrolyte, J. Colloid Sci. 17 (1962) 325-333.

161. P. Becher, Nonionic surface-active compounds, V, Effect of electrolyte - Addendum, J. Colloid Sci. 18 (1963) 196-197.

162. K. Kuriyama, Temperature dependence of micellar weight of nonionic surfactant in the presence of various additives, Part 2, Addition of sodium chloride and calcium chloride, Kolloid Z.Z. Polym. 181 (1962) 144-149.

163. M.J. Schick, Surface films of nonionic detergents, I, Surface tension study, J. Colloid Sci. 17 (1962) 801-813.

164. M.J. Schick, Physical chemistry of nonionic detergents, J. Am. Oil Chem. Soc. 40 (1963) 680-687.

165. F.E. Bailey and J.V. Koleske, Configuration and hydrodynamic properties of the polyoxyethylene chain in solution, in M.J. Schick (Ed.), Nonionic Surfactants, Dekker, New York, 1967, pp. 794-822.

166. H. Saito and K. Shinoda, The solubilization of hydrocarbons in aqueous solutions of nonionic surfactants, J. Colloid Interface Sci. 24 (1967) 10-15.

167. W.U. Malik and S.M. Saleem, Effect of additives on the critical micelle concentration of some polyethoxylated nonionic detergents, J. Am. Oil Chem. Soc. 45 (1968) 670-672.

168. A. Doren and J. Goldfarb, Electrolyte effects on micellar solutions of nonionic detergents, J. Colloid Interface Sci. 32 (1970) 67-72.

169. K. Shinoda and H. Takeda, The effect of added salts in water on the hydrophile-lipophile balance of nonionic surfactants: the effect of added salts on the phase inversion temperature of emulsions, J. Colloid Interface Sci. 32 (1970) 642-646.

170. A. Ray and G. Nemethy, Effects of ionic protein denaturants on micelle formation by nonionic detergents, J. Am. Chem. Soc. 93

(1971) 6787-6793.

171.   F. Popescu and I. Ana, Influenta electrolitilor asupra temperaturii de iesire din solutie a agentilor activi de suprafata (AAS) neionici, Pet. Gaze 24 (1973) 88-90.

172.   H. Schott, Salting in of nonionic surfactants by complexation with inorganic salts, J. Colloid Interface Sci. 43 (1973) 150-155.

173.   K. Deguchi and K. Meguro, The effects of inorganic salts and urea on the micellar structure of nonionic surfactant, J. Colloid Interface Sci. 50 (1975) 223-227.

174.   A.T. Florence, F. Madsen and F. Puisieux, Emulsion stabilization by nonionic surfactants: the relevance of surfactant cloud point, J. Pharm. Pharmacol. 27 (1975) 385-394.

175.   H. Schott and S.K. Han, Effect of inorganic additives on solutions of nonionic surfactants, II, J. Pharm. Sci. 64 (1975) 658-664.

176.   H. Schott and S.K. Han, Effect of inorganic additives on solutions of nonionic surfactants, III, CMCs and surface properties, J. Pharm. Sci. 65 (1976) 975-978.

177.   H. Schott and S.K. Han, Effect of inorganic additives on solutions of nonionic surfactants, IV, Kraft points, J. Pharm. Sci. 65 (1976) 979-981.

178.   H. Schott and S.K. Han, Effect of symmetrical tetraalkylam- monium salts on cloud point of nonionic surfactants, J. Pharm. Sci. 66 (1977) 165-168.

179.   N. Nishikido and R. Matuura, The effect of added inorganic salts on the micelle formation of nonionic surfactants in aqueous solutions, Bull. Chem. Soc. Japan 50 (1977) 1690-1694.

180.   J.S. D'Arrigo, Structural features of the nonionic surfactants stabilizing long-lived bubble nuclei, J. Chem. Phys. 72 (1980) 5133-5138.

181.   J.S. D'Arrigo, Screening of membrane surface charges by divalent cations: an atomic representation, Am. J. Physiol. 235 (1978) 109-117.

182.   M. Gouy, Sur la constitution de le charge electrique a la surface d'un electrolyte, C. R. Hebd. Seanc. Acad. Sci. Paris 149 (1909) 654-657.

183.   D.L. Chapman, A contribution to the theory of electrocapillarity, Philos. Mag. 25 (1913) 475-481.

184.   D.C. Grahame, The electrical double layer and the theory of electrocapillarity, Chem. Rev. 41 (1947) 441-501.

274

185. D.C. Grahame, Diffuse double layer theory for electrolytes of unsymmetrical valence types, J. Chem. Phys. 21 (1953) 1054-1060.

186. J.T.G. Overbeek, Electrochemistry of the double layer, in H.R. Kruyt (Ed.), Colloid Science, Vol. 1, Elsevier, Amsterdam, 1952, pp.115-193.

187. J.S. D'Arrigo, Possible screening of surface charges on crayfish axons by polyvalent metal ions, J. Physiol. 231 (1973) 117-128.

188. J.S. D'Arrigo, Axonal surface charges: binding or screening by divalent cations governed by external pH, J. Physiol. 243 (1974) 757-764.

189. J.S. D'Arrigo, Strontium ions and membranes: screening versus binding at charges surfaces, in S.C. Skoryna (Ed.), Handbook of Stable Strontium, Plenum, New York, 1981, pp. 167-182.

190. E.A. Hemmingsen, Cavitation in gas-supersaturated solutions, J. Appl. Phys. 46 (1975) 213-218.

191. W.A. Gerth and E.A. Hemmingsen, Gas supersaturation thresholds for spontaneous cavitation in water with gas equilibration pressures up to 570 atm, Z. Naturforsch. A31 (1976) 1711-1716.

192. E.A. Hemmingsen, Spontaneous formation of bubbles in gas-supersaturated water, Nature 267 (1977) 141-142.

193. E.A. Hemmingsen, Effects of surfactants and electrolytes on the nucleation of bubbles in gas-supersaturated solutions, Z. Naturforsch. A33 (1978) 164-171.

194. A.W. Adamson, Physical Chemistry of Surfaces, 3rd ed., Wiley, New York, 1976, 698 pp.

195. E.E. Schrier and E.B. Schrier, The salting-out behavior of amides and its relation to the denaturation of proteins by salts, J. Phys. Chem. 71 (1967) 1851-1860.

196. J.J. Kabara, A.J. Conley and J.P. Truant, Relationship of chemical structure and antimicrobial activity of alkyl amides and amines, Antimicrob. Agents Chemother. 2 (1972) 492-498.

197. J.J. Kabara and G.V. Haitsma, Aminimides, II, Antimicrobial effect of short chain fatty acid derivatives, J. Am. Oil Chem. Soc. 52 (1975) 444-447.

198. H.R. Chipalkatti, C.H. Giles and D.G.M. Vallance, Adsorption at organic surfaces, Part I, Adsorption of organic compounds by polyamide and protein fibers from aqueous and nonaqueous solutions, J. Chem. Soc. (1954) 4375-4390.

199. F.M. Arshid, C.H. Giles, S.K. Jain and A.S.A. Hassan, Studies in hydrogen-bond formation, Part III, The reactivity of amines, amides, and azo-compounds in aqueous and nonaqueous solutions, J. Chem. Soc. (1956) 72-75.

200. D.R. Robinson and W.P. Jencks, The effect of concentrated salt solutions on the activity coefficient of acetyltetraglycine ethyl ester, J. Am. Chem. Soc. 87 (1965) 2470-2479.

201. W.C. Griffin, Calculation of HLB values of nonionic surfactants, J. Soc. Cosmet. Chem. 5 (1954) 249-256.

202. H.L. Greenwald, G.L. Brown and M.N. Fineman, Determination of the hydrophile-lipophile character of surface active agents and oils by a water titration, Anal. Chem. 28 (1956) 1693-1697.

203. C.D. Moore and M. Bell, Nonionic surface-active agents and their utilization in cosmetic preparations, Soap Perfum. Cosmet. 29 (1956) 893-896.

204. J.T. Davies, A quantitative kinetic theory of emulsion type, I, Physical chemistry of the emulsifying agent, in J.H. Schulman (Ed.), Proceedings of the 2nd International Congress of Surface Activity, Butterworths, London, 1957, pp. 426-438.

205. H. Schott, Hydrophilic-lipophilic balance and distribution coefficients of nonionic surfactants, J. Pharm. Sci. 60 (1971) 648-649.

206. R.W. Egan, M.A. Jones and A.L. Lehninger, Hydrophile-lipophile balance and critical micelle concentration as key factors influencing surfactant disruption of mitochondrial membranes, J. Biol. Chem. 251 (1976) 4442-4447.

207. C. Pacifico and M.E. Ionescu, Surfactant survey, J. Am. Oil Chem. Soc. 34 (1957) 203-210.

208. C.E. Stevens, Nonionic surfactants, J. Am. Oil Chem. Soc. 34 (1957) 181-185.

209. W.U. Malik and O.P. Jhamb, Critical micelle concentration of polyoxyethylated nonionic surfactants and the effect of additives, Kolloid Z.Z. Polym. 242 (1970) 1209-1211.

210. A.R. Trussell and M.D. Umphres, An overview of the analysis of trace organics in water, J. Am. Water Works Assoc. 70 (1978) 595-603.

211. G.A. Junk, Organic contaminants in water, in F. Coulston and E. Mrak (Eds.), Water Quality, Proceedings of an International Forum, Academic Press, New York, 1977, pp.137-141.

212. M.J. Rosen and H.A. Goldsmith, Systematic Analysis of Surface-

Active Agents, 2nd ed., Wiley-Interscience, London, 1972, 591 pp.

213. G.F. Longman, The Analysis of Detergents and Detergent Products, Wiley-Interscience, London, 1975, 587 pp.

214. P.M. Williams and A. Zirino, Scavenging of "dissolved" organic matter from sea water with hydrated metal oxides, Nature 204 (1964) 462-464.

215. E.K. Duursma, The dissolved organic constituents of sea water, in J.P. Riley and G. Skirrow (Eds.), Chemical Oceanography, Academic Press, New York, 1965, pp. 433-475.

216. P.J. Wangersky, The organic chemistry of sea water, Am. Sci. 53 (1965) 358-374.

217. K. Kalle, The problem of the Gelbstoff in the sea, Oceanogr. Mar. Biol. Ann. Rev. 4 (1966) 91-104.

218. R.T. Barber, Dissolved organic carbon from deep waters resists microbial oxidation, Nature 220 (1968) 274-275.

219. K.M. Khaylov, Dissolved organic macromolecules in sea water, Geokhimiya 5 (1968) 595-603.

220. B. Skopintsev, On the age of stable organic matter-aquatic humus in oceanic waters, in D. Dyrssen and D. Jagner (Eds.), The Changing Chemistry of the Oceans, Wiley, New York, 1972, pp. 205-207.

221. T.R. Parsons and J.D.H. Strickland, Oceanic detritus, Science 136 (1962) 313-314.

222. E.R. Baylor and W.H. Sutcliffe, Dissolved organic matter in seawater as a source of particulate food, Limnol. Oceanogr. 8 (1963) 369-371.

223. A. Nissenbaum and I.R. Kaplan, Chemical and isotopic evidence for the in situ origin of marine humic substances, Limnol. Oceanogr. 17 (1972) 570-582.

224. D.H. Stuermer and G.R. Harvey, Humic substances from seawater, Nature 250 (1974) 480-481.

225. R.A. Kerr and J.G. Quinn, Chemical studies on the dissolved organic matter in seawater: isolation and fractionation, Deep-Sea Res. 22 (1975) 107-116.

226. G.I. Loeb and R.A. Neihof, Marine conditioning films, Adv. Chem. Series 145 (1975) 319-335.

227. D.H. Stuermer and G.R. Harvey, The isolation of humic substances and alcohol-soluble organic matter from seawater, Deep-Sea Res. 24 (1977) 303-309.

228. D.C. Blanchard and E.J. Hoffman, Control of jet drop dynamics by organic material in seawater, J. Geophys. Res. 83 (1978) 6317-6191.

229. W.D. Garrett, Reply to comment on retardation of water drop evaporation with monomolecular surface films, J. Atmos. Sci. 29 (1972) 786-787.

230. W.D. Garrett, Comment on "Organic particle and aggregate formation resulting from the dissolution of bubbles in seawater", Limnol. Oceanogr. 26 (1981) 989-992.

231. J.S. D'Arrigo, Aromatic proteinaceous surfactants stabilize long-lived gas microbubbles from natural sources, J. Chem. Phys. 75 (1981) 962-968.

232. R.J. Block, E.L. Durrum and G. Zwelg, A Manual of Paper Chromatography and Paper Electrophoresis, Academic Press, New York, 1958.

233. J.P. Greenstein and M. Winitz, Chemistry of the Amino Acids, Wiley, New York, 1961.

234. C. Araki and K. Arai, Studies on the chemical constitution of agar-agar, XX, Isolation of a tetrasaccharide by enzymatic hydrolysis of agar-agar, Bull. Chem. Soc. Japan 30 (1957) 287-293.

235. C. Araki and S. Hirase, Chemical constitution of agar-agar, XXI, Reinvestigation of methylated agarose of Gelidium amansii, Bull. Chem. Soc. Japan 33 (1960) 291-295.

236. S. Hjerten, New method for preparation of agarose for gel electrophoresis, Biochem. Biophys. Acta 62 (1962) 445-449.

237. B. Russell, T.H. Mead and A. Polson, Method for preparing agarose, Biochem. Biophys. Acta 86 (1964) 169-174.

238. J.C. Hegenauer and G.W. Nace, Improved method for preparing agarose, Biochem. Biophys. Acta 111 (1965) 334-336.

239. S.J. Barteling, Preparation of agarose, Clin. Chem. 15 (1969) 1002-1005.

240. G.G. Allan, P.G. Johnson, Y.Z. Lai and K.V. Sarkanen, Marine polymers, Part I, A new procedure for the fractionation of agar, Carbohyd. Res. 17 (1971) 234-236.

241. E.C. Shorey, Organic nitrogen in Hawaiian soils, Annu. Rep. Hawaii Agr. Exp. Sta. (1906) 37-59.

242. O. Schreiner and E.C. Shorey, The isolation of harmful organic substances from soils, U.S. Dep. Agr. Bull. 53 (1909) 1-45.

243. S.L. Jodidi, The chemical nature of the organic nitrogen in the

278

soil, J. Am. Chem. Soc. 33 (1911) 1226-1241.

244.   S.L. Jodidi, The chemical nature of the organic nitrogen in the soil, Second paper, J. Am. Chem. Soc. 34 (1912) 94-99.

245.   W.P. Kelley, The organic nitrogen of Hawaiian soils, I, The products of acid hydrolysis, J. Am. Chem. Soc. 36 (1914) 429-434.

246.   W.P. Kelley, The organic nitrogen of Hawaiian soils, II, The effects of heat on soil nitrogen, J. Am. Chem. Soc. 36 (1914) 434-438.

247.   W.P. Kelley and A.R. Thompson, The organic nitrogen of Hawaiian soils, III, The nitrogen of humus, J. Am. Chem. Soc. 36 (1914) 438-444.

248.   R.S. Potter and R.S. Snyder, Amino acid nitrogen of soils and the chemical groups of amino acids in the hydrolyzed soil and their humic acids, J. Am. Chem. Soc. 37 (1915) 2219-2224.

249.   E.H. Waletrs, The presence of proteoses and peptones in soils, J. Ind. Engng. Chem. 7 (1915) 860-863.

250.   E.C. Lathrop, Protein decomposition in soils, Soil Sci. 1 (1916) 509-532.

251.   W.L. Davies, The proteins of different types of peat soils, J. Agr. Sci. 18 (1928) 682-690.

252.   R.P. Hobson and H.J. Page, Studies on the carbon and nitrogen cycles in the soil, VII, The nature of the organic nitrogen compounds of the soil: "humic" nitrogen, J. Agr. Sci. 22 (1932) 497-515.

253.   R.P. Hobson and H.J. Page, Studies on the carbon and nitrogen cycles in the soil, VIII, The nature of the organic nitrogen compounds of the soil: "non-humic" nitrogen, J. Agr. Sci. 22 (1932) 516-526.

254.   R.T. Kojima, Soil organic nitrogen, I, Nature of the organic nitrogen in a muck soil from Geneva, New York, Soil Sci. 64 (1947) 157-165.

255.   R.T. Kojima, Soil organic nitrogen, II, Some studies on the amino acids of protein material in a muck soil from Genava, New York, Soil Sci. 64 (1947) 245-252.

256.   J.M. Bremner, The amino-acid composition of the protein material in soil, Biochem. J. 47 (1950) 538-542.

257.   J.M. Bremner, A review of recent work on soil organic matter, Part I, J. Soil Sci. 2 (1951) 67-82.

258.   F.J. Stevenson, Ion exchange chromatography of the amino acids

in soil hydrolysates, Soil Sci. Soc. Am. Proc. 18 (1954) 373-377.

259. F.J. Stevenson, Isolation and identification of some amino compounds in soils, Soil Sci. Soc. Am. Proc. 20 (1956) 201-204.

260. F.J. Stevenson, Effect of some long-time rotations on the amino acid composition of the soil, Soil Sci. Soc. Am. Proc. 20 (1956) 204-208.

261. F.J. Sowden, The forms of nitrogen in the organic matter of different horizons of soil profiles, Can. J. Soil Sci. 38 (1958) 147-154.

262. J.L. Young and J.L. Mortensen, Soil nitrogen complexes, I, Chromatography of amino compounds in soil hydrolysates, Ohio Agr. Exp. Sta. Res. Circ. 61 (1958) 1-18.

263. J.M. Bremner, Organic nitrogen in soils, in W.V. Bartholomew and F.E. Clark (Eds.), Soil Nitrogen, American Society of Agronomy, Madison, 1965, pp. 93-120.

264. A.M. Briones, Nature and distribution of organic nitrogen in tropical soils, Ph.D. Thesis, University of Hawaii, 1969.

265. P. Simonart, L. Batistic and J. Mayaudon, Isolation of protein from humic acid extracted from soil, Plant Soil 27 (1967) 153-161.

266. V.O. Biederbeck and E.A. Paul, Fractionation of soil humate with phenolic solvents and purification of the nitrogen-rich portion with polyvinylpyrrolidone, Soil Sci. 115 (1973) 357-366.

267. J.P. Martin, A.A. Parsa and K. Haider, Influence of intimate association with humic polymers on biodegradation of [^{14}C] labeled organic substrates in soil, Soil Biol. Biochem. 10 (1978) 483-486.

268. P.A. Cranwell and R.D. Haworth, The chemical nature of humic acids, in D. Povoledo and H.L. Golterman (Eds.), Humic Substances: Their Structure and Function in the Biosphere, Centre of Agricultural Publishing and Documentation, Wageningen, 1975, pp. 13-18.

269. L. Verma, J.P. Martin and K. Haider, Decomposition of carbon-14-labeled proteins, peptides, and amino acids; free and complexed with humic polymers, Soil Sci. Soc. Am. Proc. 39 (1975) 279-284.

270. A. White, P. Handler and E.L. Smith, Principles of Biochemistry, 3rd ed., McGraw-Hill, New York, 1964, 1106 pp.

271. A. Knowles and S. Gurnani, A study of the methylene blue-sensitized oxidation of amino acids, Photochem. Photobiol. 16

(1972) 95-108.

272. L. Weil, W.G. Gordon and A.R. Buchert, Photooxidation of amino acids in the presence of methylene blue, Arch. Biochem. Biophys. 33 (1951) 90-109.

273. L. Weil, On the mechanism of the photooxidation of amino acids sensitized by methylene blue, Arch. Biochem. Biophys. 110 (1965) 57-68.

274. R.D. Guy, D.R. Narine and S. DeSilva, Organo-cation speciation, I, A comparison of the interactions of methylene blue and paraquat with bentonite and humic acid, Can. J. Chem. 58 (1980) 547-554.

275. L. Weil, A.R. Buchert and J. Maher, Photooxidation of crystalline lysozyme in the presence of methylene blue and its relation to enzymatic activity, Arch. Biochem. Biophys. 40 (1952) 245-252.

276. L. Weil, S. James and A.R. Buchert, Photooxidation of crystalline chymotrypsin in the presence of methylene blue, Arch. Biochem. Biophys. 46 (1953) 266-278.

277. J.D. Spikes and R. Livingston, The molecular biology of photodynamic action: sensitized photoautoxidations in biological systems, in L.G. Augenstein, R. Mason and M. Zelle (Eds.), Advances in Radiation Biology, Vol. 3, Academic Press, New York, 1969, pp. 29-110.

278. H.R. Horton, H. Kelly and D.E. Koshland, Environmentally sensitive protein reagents: 2-methoxy-5-nitrobenzyl bromide, J. Biol. Chem. 240 (1965) 722-724.

279. T.E. Barman and D.E. Koshland, A colorimetric procedure for the quantitative determination of tryptophan residues in proteins, J. Biol. Chem. 242 (1967) 5771-5776.

280. B.L. Vallee and J.F. Riordan, Chemical approaches to the properties of active sites of enzymes, Annu. Rev. Biochem. 38 (1969) 733-784.

281. G.R. Stark, Recent developments in chemical modification and sequential degradation of proteins, Adv. Protein Chem. 24 (1970) 261-308.

282. Y.D. Karkhanis, D.J. Carlo and J. Zeltner, A simplified procedure to determine tryptophan residues in proteins, Anal. Biochem. 69 (1975) 55-60.

283. E. Gorter, Spreading in a monomolecular film, A method for studying biologic problems, Am. J. Dis. Child 47 (1934) 945-

957.

284. H. Neurath, The influence of denaturation on the spreading of proteins on a water surface, J. Phys. Chem. 40 (1936) 361-368.

285. H. Neurath and H.B. Bull, The surface activity of proteins, Chem. Rev. 23 (1938) 391-435.

286. K.G.A. Pankhurst, Foam formation and foam stability: the effect of the adsorbed layer, Trans. Faraday Soc. 37 (1941) 496-505.

287. J.M. Perri, Jr. and F. Hazel, Lime hydrolysis of soybean protein, Ind. Engng. Chem. 38 (1946) 549-554.

288. J.M. Perri, Jr. and F. Hazel, Effect of electrolytes on the foaming capacity of alpha soybean protein dispersions, J. Phys. Colloid Chem. 51 (1947) 661-666.

289. S.J. Singer, Note on an equation of state for linear macromolecules in monolayers, J. Chem. Phys. 16 (1948) 872-876.

290. S.G. Davis, C.R. Fellers and W.B. Esselen, Jr., Foam fractionation procedures in the isolation of fruit proteins, Food Tech. 3 (1949) 198-201.

291. W.C. Thuman, A.G. Brown and J.W. McBain, Studies of protein foams obtained by bubbling, J. Am. Chem. Soc. 71 (1949) 3129-3135.

292. C.W.N. Cumper and A.E. Alexander, The surface chemistry of proteins, Trans. Faraday Soc. 46 (1950) 235-253.

293. C.W.N. Cumper, The stabilization of foams by proteins, Trans. Faraday Soc. 49 (1953) 1360-1369.

294. J.T. Davies, Some factors influencing the orientation of ε-amino groups in monolayers of proteins and amino acid polymers, Biochem. J. 56 (1954) 509-513.

295. R.J. Goldacre, Surface films, their collapse on compression, the shapes and sizes of cells and the origin of life, in J.F. Danielli, K.G.A. Pankhurst and A.C. Riddiford (Eds.), Surface Phenomena in Chemistry and Biology, Pergamon, New York, 1958, pp. 278-298.

296. B.R. Malcolm, Conformation of synthetic polypeptide and protein monolayers at interfaces, Nature 195 (1962) 901-902.

297. R. Nakamura, Studies on the foaming property of the chicken egg white, Part VII, On the foaminess of the denatured ovalbumin, Agr. Biol. Chem. 28 (1964) 403-407.

298. J.F.T. Oldfield and J.V. Dutton, Surface active constituents in beet sugar crystallization, Int. Sugar J. 70 (1968) 7-9.

299. B.R. Malcolm, Molecular structure and deuterium exchange in

282

monolayers of synthetic polypeptides, Proc. R. Soc. London Ser. A 305 (1968) 363-385.

300. D.D. Jones and M. Jost, Isolation and chemical characterization of gas-vacuole membranes from *Microcystis aeruginosa*, Elenkin. Arch. Mikrobiol. 70 (1970) 43-64.

301. J.J. Bikerman, Foams, Springer, New York, 1973.

302. R.J. King, D.J. Klass, E.G. Gikas and J.A. Clements, Isolation of apoproteins from canine surface active material, Am. J. Physiol. 224 (1973) 788-795.

303. B. Heard, R. White and E. King, Effects of serum proteins on pulmonary surfactants *in vitro*: evidence of a protective mechanism in pulmonary edema, Chest 66 (1974) 13S-16S.

304. K. Kalischewski, W. Bumbullis and K. Schugerl, Foam behavior of biological media, I, Protein foams, Eur. J. Appl. Microbiol. Biotechnol. 7 (1979) 21-31.

305. K. Kalischewski and K. Schugerl, Investigation of protein foams obtained by bubbling, Colloid Polym. Sci. 257 (1979) 1099-1110.

306. R.D. Waniska and J.E. Ninsella, Foaming properties of proteins: evaluation of a column aeration apparatus using ovalbumin, J. Food Sci. 44 (1979) 1398-1402.

307. M.C. Phillips, Protein conformation at liquid interfaces and its role in stabilizing emulsions and foams, Food Tech. 35 (1981) 50-57.

308. M.A. Passero, R.W. Tye, K.H. Kilburn and W.S. Lynn, Isolation and characterization of two glycoproteins from patients with alveolar proteinosis, Proc. Nat. Acad. Sci. 70 (1973) 973-976.

309. S.N. Bhattacharyya, M.A. Passero, R.P. DiAugustine and W.S. Lynn, Isolation and characterization of two hydroxyproline-containing glycoproteins from normal animal lung lavage and lamellar bodies, J. Clin. Invest. 55 (1975) 914-920.

310. S.N. Bhattacharyya, M.C. Rose, M.G. Lynn, C. MacLeod, M. Alberts and W.S. Lynn, Isolation and characterization of a unique glycoprotein from lavage of chicken lungs and lamellar organelles, Am. Rev. Resp. Dis. 114 (1976) 843-850.

311. S.N. Bhattacharyya. S. Sahu and W.S. Lynn, Structural studies on a glycoprotein isolated from alveoli of patients with alveolar proteinosis, Biochem. Biophys. Acta 427 (1976) 91-106.

312. S.N. Bhattacharyya and W.S. Lynn, Studies on structural relationship between two glycoproteins isolated from alveoli of patients with alveolar proteinosis, Biochem. Biophys. Acta 494

(1977) 150-161.

313. S.N. Bhattacharyya and W.S. Lynn, Structural studies on the oligosaccharides of a glycoprotein isolated from alveoli of patients with alveolar proteinosis, J. Biol. Chem. 252 (1977) 1172-1180.

314. R.J. King, H. Martin, D. Mitts and F.M. Holmstrom, Metabolism of the apoproteins in pulmonary surfactant, J. Appl. Physiol. 42 (1977) 483-491.

315. S.N. Bhattacharyya and W.S. Lynn, Isolation and characterization of a pulmonary glycoprotein from human amniotic fluid, Biochem. Biophys. Acta 537 (1978) 329-335.

316. S.N. Bhattacharyya and W.S. Lynn, Structural characterization of a glycoprotein isolated from alveoli of patients with alveolar proteinosis, J. Biol. Chem. 254 (1979) 5191-5198.

317. S.L. Katyal, and G. Singh, An immunologic study of the apoproteins of rat lung surfactant, Lab. Invest. 40 (1979) 562-567.

318. S.N. Bhattacharyya and W.S. Lynn, Structural studies on a collagen-like glycoprotein isolated from lung lavage of normal animal, Biochem. Biophys. Acta 626 (1980) 451-458.

319. S.L. Katyal and G. Singh, Analysis of pulmonary surfactant apoproteins by electrophoresis, Biochem. Biophys. Acta 670 (1981) 323-331.

320. V.P. Lehto, I. Kantola, T. Tervo and L.A. Laitinen, Ruthenium red staining of blood-bubble interface in acute decompression sickness in rats, Undersea Biomed. Res. 8 (1981) 101-111.

321. G. Ogner and M. Schnitzer, Chemistry of fulvic acid, a soil humic fraction, and its relation to lignin, Can. J. Chem. 49 (1971) 1053-1063.

322. J.S. D'Arrigo, Biological surfactants stabilizing natural micro-bubbles in aqueous media, Adv. Colloid Interface Sci. 19 (1983) 253-307.

323. G.Y. Barabanova, V.P. Il'in, Y.L. Levkovskii and A.V. Chalov, Relationship between the strength and size of cavitation nuclei, Sov. Phys. Acoust. 27 (1981) 25-28.

324. V.P. Il'in, Y.L. Levkovskii and A.V. Chalov, Influence of the concentration and distribution of cavitation nuclei on the inception and noise of bubble cavitation, Sov. Phys. Acoust. 27 (1981) 220-222.

325. L.P. Kravtsova and E.S. Chistyakov, Influence of the insonifica-

284

tion time of a liquid on the threshold pressures for gaseous cavitation, Sov. Phys. Acoust. 27 (1981) 249-250.

326. S. Moore and W.H. Stein, Chromatographic determination of amino acids by the use of automatic recording equipment, in S.P. Colowick and N.O. Kaplan(Eds.), Methods in Enzymology, Vol. VI, Academic Press, New York, 1963, pp. 819-831.

327. N.W. Downer, N.C. Robinson and R.A. Capaldi, Components of the mitochondrial inner membrane, 3, Characterization of a seventh different subunit of beef heart cytochrome c oxidase, Similarities between the beef heart enzyme and that from other species, Biochemistry 15 (1976) 2930-2936.

328. R.T. Swank and K.D. Munkres, Molecular weight analysis of oligopeptides by electrophoresis in polyacrylamide gel with sodium dodecyl sulfate, Anal. Biochem. 39 (1971) 462-477.

329. H. Glossmann and D.M. Neville, Glycoproteins of cell surfaces, Comparative study of three different cell surfaces of the rat, J. Biol. Chem. 246 (1971) 6339-6346.

330. K. Weber and M. Osborn, Reliability of molecular weight determinations by dodecyl sulfate-polyacrylamide-gel electrophoresis, J. Biol. Chem. 244 (1969) 4406-4412.

331. A. Neuberger and R.D. Marshall, Methods for the qualitative and quantitative analysis of the component sugars, in A. Gottschalk (Ed.), Glycoproteins: Their Composition, Structure and Function, Elsevier, Amsterdam, 1966, pp. 190-234.

332. B.S. Hartley, Strategy and tactics in protein chemistry, First BDH lecture, Biochem. J. 119 (1970) 805-822.

333. D. Morse and B.L. Horecker, Thin-layer chromatographic separation of DNS-amino acids, Anal. Biochem. 14 (1966) 429-433.

334. G.E. Tarr, Improved manual sequencing methods, in C.W. Hirs and S.N. Timasheff (Eds.), Methods in Enzymology, Vol. XLVII, Enzyme Structure, Part E, Academic Press, New York, 1977, pp. 335-357.

335. J.P. Segrest and R.L. Jackson, Molecular weight determination of glycoproteins by polyacrylamide gel electrophoresis in sodium dodecyl sulfate, in V. Ginsburg (Ed.), Methods in Enzymology, Vol. XXVIII, Complex Carbohydrates, Part B, Academic Press, New York, 1972, pp. 54-63.

336. K. Weber, J.R. Pringle and M. Osborn, Measurement of molecular weights by electrophoresis on SDS-acrylamide gel, in C.H.W. Hirs and S.N. Timasheff (Eds.), Methods in

Enzymology, Vol. XXVI, Enzyme Structure, Part C, Academic Press, New York, 1972, pp. 3-27.

337. Anonymous, Gel Filtration, Theory and Practice, Pharmacia, Uppsala, 1979.

338. K.G. Mann and W.W. Fish, Protein polypeptide chain molecular weights by gel chromatography in guanidinium chloride, in C.H.W. Hirs and S.N. Timasheff (Eds.), Methods in Enzymology, Vol. XXVI, Enzyme Structure, Part C, Academic Press, New York, 1972, pp. 28-42.

339. P. Andrews, The gel-filtration behavior of proteins related to their molecular weights over a wide range, Biochem. J. 96 (1965) 595-606.

340. Anonymous, Sephadex LH-20 Chromatography in Organic Solvents, Pharmacia, Uppsala, 1978.

341. R. Montgomery, Heterogeneity of the carbohydrate groups of glycoproteins, in A. Gottschalk (Ed.), Glycoproteins: Their Composition, Structure, and Function - Part A, 2nd ed., Elsevire, Amsterdam, 1972, pp. 518-528.

342. R.E. Baier, D.W. Goupil, S. Perlmutter and R. King, Dominant chemical composition of sea-surface films, natural slicks, and foams, J. Rech. Atmos. 8 (1974) 571-600.

343. V.P. Lehto and L.A. Laitinen, Scanning and transmission electron microscopy of the blood-bubble interface in decompressed rats, Aviat. Space Environ. Med. 50 (1979) 803-807.

344. P.S. Russo, F.D. Blum, J.D. Ipsen, Y.J. Abul-Hajj and W.G. Miller, The surface activity of the phytotoxin cerato-ulmin, Can. J. Botan. 60 (1982) 1414-1422.

345. P.S. Russo, F.D. Blum, J.D. Ipsen, Y.J. Abul-Hajj and W.G. Miller, Solubility and surface activity of the *Ceratocystis ulmi* toxin cerato-ulmin, Physiol. Plant Path. 19 (1981) 113-126.

346. S. Takai and W.C. Richards, Cerato-ulmin, a wilting toxin of *Ceratocystis-ulmi*, Isolation and some properties of cerato-ulmin from culture of *C.-ulmi*, Phytopath. Z. 91 (1978) 129-146.

347. K.J. Stevenson, J.A. Slater and S. Takai, Cerato-ulmin, wilting toxin of Dutch Elm disease fungus, Phytochem. 18 (1979) 235-238.

348. R.P. Quintana, Surface chemistry of dental integuments, Adv. Chem. Series 145 (1975) 290-299.

349. T. Sonju and G. Rolla, Chemical analysis of the acquired pellicle

formed in two hours on cleaned human teeth in vivo, Rate of formation and amino acid analysis, Caries Res. 7 (1973) 30-38.

350. C.W. Mayhall, Concerning the composition and source of the acquired enamel pellicle of human teeth, Arch. Oral Biol. 15 (1970) 1327-1341,

351. K.C. Beck, J.H. Reuter and E.M. Perdue, Organic and inorganic geochemistry of some coastal plain rivers of the southeastern United States, Geochim. Cosmochim. Acta 38 (1974) 341-364.

352. E. Peake, B.L. Baker and G.W. Hodgson, Hydrogeochemistry of surface waters of the Mackenzie River drainage basin, Canada, 2, Contribution of amino acids, hydrocarbons and chlorins to Beaufort Sea by Mackenzie River system, Geochim. Cosmochim. Acta 36 (1972) 867-883.

353. C.R. Lytle and E.M. Perdue, Free, proteinaceous, and humic-bound amino acids in river water containing high concentrations of aquatic humus, Environ. Sci. Tech. 15 (1981) 224-228.

354. C.M. Preston, S.P. Mathur and B.S. Rauthan, The distribution of copper, amino compounds, and humus fractions in organic soils of different copper content, Soil Sci. 131 (1981) 344-352.

355. R.F. Christman and M. Ghassemi, Chemical nature of organic color in water, J. Amer. Water Works Assoc. 55 (1966) 723-741.

356. J.S. D'Arrigo, C. Saiz-Jimenez and N.S. Reimer, Geochemical properties and biochemical composition of the surfactant mixture surrounding natural microbubbles in aqueous media, J. Colloid Interface Sci. 100 (1984) 96-105.

357. D.E. Foote (Ed.), Soil Survey of Islands of Kauai, Oahu, Maui, Molokai, and Lanai, State of Hawaii, U.S. Soil Conservation Service, U.S. Government Printing Office, Wash. DC, 1972, 232 pp.

358. K. Haider, B.R. Nagar, C. Saiz-Jimenez, H.L.C. Meuzelaar and J.P. Martin, Studies on soil humic compounds, fungal melanins and model polymers by pyrolysis mass spectrometry, in Soil Organic Matter Studies, Vol. II, IAEA, Vienna, 1977, pp. 213-220.

359. C. Saiz-Jimenez, K. Haider and H.L.C. Meuzelaar, Comparisons of soil organic matter and its fractions by pyrolysis mass spectrometry, Geoderma 22 (1979) 25-37.

360. J. Haverkamp, H.L.C. Meuzelaar, E.C. Beuvery, P.M. Boonekamp and R.H. Tiesjema, Characterization of *Neisseria meningitidis* capsular polysaccharides containing sialic acid by

pyrolysis mass spectrometry, Anal. Biochem. 104 (1980) 407-418.

361.   J.S. D'Arrigo, Surface properties of microbubble-surfactant monolayers at the air/water interface, J. Colloid Interface Sci. 100 (1984) 106-111.

362.   F. Martin and C. Saiz-Jimenez, A water soluble fraction of humic acids from vertisols, Z. Pflanzenernaer. Bodenkd. 135 (1973) 58-67.

363.   M. Schnitzer, Humic substances - chemistry and reactions, in M. Schnitzer and S.U. Khan (Eds.), Soil Organic Matter, Elsevier, Amsterdam, 1978, pp.1-64.

364.   J.F. Dormaar, Water-soluble constituents in defrost water of Black Chernozemic A horizons during Spring thaw, Can. J. Soil Sci. 58 (1978) 135-144.

365.   H.L.C. Meuzelaar, P.G. Kistemaker and M.A. Posthumus, Recent advances in pyrolysis mass spectrometry of complex biological materials, Biomed. Mass Spectrom. 1 (1974) 312-319.

366.   F. Martin, C. Saiz-Jimenez and A. Cert, Pyrolysis gas chromatography-mass spectrometry of soil humic fractions, 2, High boiling-point compounds, Soil Sci. Soc. Am. Proc. 43 (1979) 309-312.

367.   J.C. Vedy and S. Bruckert, Seasonal evolution of soluble organic compounds in relation to different biochemical humification processes, Pedologie 20 (1970) 135-152.

368.   H.J. Dawson, F.C. Ugolini, B.F. Hrutfiord and J. Zackara, Role of soluble organics in soil processes of a podzol, Central Cascades, Washington, Soil Sci. 126 (1978) 290-296.

369.   S.U. Khan, Organic matter association with soluble salts in the water extract of a black solonetz soil, Soil Sci. 109 (1970) 227-228.

370.   R. Candler and K. van Cleve, A comparison of aqueous extracts from the B horizon of a birch and aspen forest in interior Alaska, Soil Sci. 134 (1982) 176-180.

371.   F. MacRitchie and A.E. Alexander, The effect of sucrose on protein films, I, Spread monolayers, J. Colloid Sci. 16 (1961) 57-61.

372.   F. MacRitchie and A.E. Alexander, The effect of sucrose on protein films, II, Adsorbed films, J. Colloid Sci. 16 (1961) 61-67.

373.   J.P. Thornber, Chlorophyll-proteins: light-harvesting and reaction center components of plants, Annu. Rev. Plant Physiol. 26 (1975)

288

127-158.

374. J.P. Thornber and J. Barber, Photosynthetic pigments and models for their organization in vivo, in J. Barber (Ed.), Photosynthesis in Relation to Model Systems, Elsevier, Amsterdam, 1979, pp. 27-70.

375. R.G. Hiller and D.J. Goodchild, Thylakoid membrane and pigment organization, in M.D. Hatch and N.K. Boardman (Eds.), The Biochemistry of Plants: A Comprehensive Treatise, Vol. 8, Photosynthesis, Academic Press, New York, 1981, pp. 1-49.

376. J.P. Thornber, J.C. Stewart, M.W.C. Hatton and J.L. Bailey, Studies on the nature of chloroplast lamellae, II, Chemical composition and further physical properties of two chlorophyll-protein complexes, Biochemistry 6 (1967) 2006-2014.

377. I.J. Ryrie and N. Fuad, Membrane adhesion in reconstituted proteoliposomes containing the light-harvesting chlorophyll a/b-protein complex: the role of charged surface groups, Arch. Biochem. Biophys. 214 (1982) 475-488.

378. K.S. Kan and J.P. Thornber, The light-harvesting chlorophyll a/b-protein complex of *Chlamydomonas reinhardii*, Plant Physiol. 57 (1976) 47-52.

379. V. Zutic and T. Legovic, A film of organic matter at the fresh-water/sea-water interface of an estuary, Nature 328 (1987) 612-614.

380. J.F. Padday, A.R. Pitt and R.M. Pashley, Menisci at a free liquid surface: surface tension from the maximum pull on a rod, J. Chem. Soc., Faraday Trans. I 71 (1975) 1919-1931.

381. G.L. Gaines, Insoluble monolayers, in M. Kerker (Ed.), Surface Chemistry of Colloids, (MTP International Review of Science), Vol. 7, University Park Press, Baltimore, 1972, pp. 1-24.

382. D.O. Shah, The biology of surfaces, in C.A. Villee, D.B. Villee and J. Zuckerman (Eds.), Respiratory Distress Syndrome, Academic Press, New York, 1973, pp.47-75.

383. R.H. Notter and P.E. Morrow, Pulmonary surfactant: a surface chemistry viewpoint, Ann. Biomed. Engng. 3 (1975) 119-159.

384. T. Fujiwara and F.H. Adams, Surfactant for hyaline membrane disease, Pediatrics 66 (1980) 795-798.

385. D.F. Waugh, Protein-protein interactions, Adv. Protein Chem. 9 (1954) 325-398.

386. H. Kistenmacher, H. Popkie and E. Clementi, Study of structure of molecular complexes, 3, Energy surface of a water molecule in

the field of a fluorine or chlorine anion, J. Chem. Phys. 58 (1973) 5627-5638.

387. L.C. Allen, A model for the hydrogen bond, Proc. Nat. Acad. Sci. USA 72 (1975) 4701-4705.

388. E.J. Bock and N.M. Frew, Static and dynamic response of natural multicomponent oceanic surface films to compression and dilation: laboratory and field observations, J. Geophys. Res. 98 (1993) 14599-14617.

389. S.J. Pogorzelski and A.D. Kogut, Properties of surfactant films on natural waters, Oceanologia 43 (2001) 223-246.

390. J.C. Marty, V. Zutic, R. Precali, A. Saliot, B. Cosovic, N. Smodlaka and G. Cauwet, Organic matter characterization in the Northern Adriatic Sea with special reference to the sea surface microlayer, Marine Chem. 25 (1988) 243-263.

391. B. Cosovic and V. Vojvodic, Adsorption behaviour of the hydrophobic fraction of organic matter in natural waters, Marine Chem. 28 (1989) 183-198.

392. N.M. Frew and R.K. Nelson, Scaling of marine microlayer film surface pressure-area isotherms using chemical attributes, J. Geophys. Res. 97 (1992) 5291-5300.

393. N.M. Frew and R.K. Nelson, Isolation of marine microlayer film surfactants for ex situ study of their surface physical and chemical properties, J. Geophys. Res. 97 (1992) 5281-5290.

394. J.S. D'Arrigo, B.C. Hammer and J.H. Bradbury, Structural analysis of the major surfactant components stabilizing natural microbubbles using ^{1}H- and ^{13}C-NMR spectroscopy, 1983 (limited distribution report).

395. C.Y. Hopkins, Nuclear magnetic resonance in fatty acids and glycerides, in R.T. Holman (Ed.), Progress in the Chemistry of Fats and other Lipids, Vol. VIII, Part 2, Pergamon, New York, 1966, pp. 213-252.

396. C.Y. Hopkins and H.J. Bernstein, Applications of proton magnetic resonance spectra in fatty acid chemistry, Can. J. Chem. 37 (1959) 775-782.

397. L.F. Johnson and J.N. Shoolery, Determination of unsaturation and average molecular weight of natural fats by nuclear magnetic resonance, Anal. Chem. 34 (1962) 1136-1139.

398. J. Cason and G.L. Lange, Nuclear magnetic resonance determination of substituent methyls in fatty acids, J. Org. Chem. 29 (1964) 2107-2108.

290

399. C.A. Glass and H.J. Dutton, Determination of beta-olefinic methyl groups in esters of fatty acids by nuclear magnetic resonance, Anal. Chem. 36 (1964) 2401-2404.

400. J.M. Purcell, S.G. Morris and H. Susi, Proton magnetic resonance spectra of unsaturated fatty acids, Anal. Chem. 38 (1966) 588-592.

401. N.F. Chamberlain, Nuclear magnetic resonance chemical shifts of oxygenated unsaturated aliphatics, Anal. Chem. 40 (1968) 1317-1325.

402. J.H. Bradbury and J.C. Collins, An approach to the structural analysis of oligosaccharides by n.m.r. spectroscopy, Carbohyd. Res. 71 (1979) 15-24.

403. C.J. Pouchert (Ed.), The Aldrich Library of NMR Spectra, 2nd ed., Aldrich Chemical, Milwaukee, 1983.

404. W.D. Garrett, Organic chemistry of natural sea surface films, in D.W. Hood (Ed.), Organic Matter in Natural Waters, Inst. Mar. Sci. (Alaska) Occas. Publ. 1, 1970, pp. 469-477.

405. K. Larsson, G. Odham and A. Sodergren, On lipid surface films on the sea, I, A simple method for sampling and studies of composition, Mar. Chem. 2 (1974) 49-57.

406. G. Odham, B. Noren, B. Norkrans, A. Sodergren and H. Lofgren, Biological and chemical aspects of the aquatic lipid surface monolayer, Prog. Chem. Fats other Lipids 16 (1978) 31-44.

407. S. Kjelleberg, B. Norkrans, H. Lofgren and K. Larsson, Surface balance study of interaction between microorganisms and a lipid monolayer at the air-water interface, Appl. Environmental Microbiol. 31 (1976) 609-611.

408. United States Navy Department, Diving Manual, U.S. Government Printing Office, Wash. DC, 1974.

409. A.E. Boycott, G.C.C. Damant and J.B. Haldane, The prevention of compressed-air illness, J. Hyg. (Camb.) 8 (1908) 342-443.

410. Department of Labor - Japan, Textbook for Divers, Chuo-rohdoh-saigai-bohshi-kyokai (Central Association to Prevent Labor Accidents), Tokyo, 1976.

411. H.V. Hempleman, Decompression theory: British practice, in P.B. Bennett and D.H. Elliott (Eds.), The Physiology and Medicine of Diving and Compressed Air Work, Williams & Wilkins Co., Baltimore, 1975, pp. 331-347.

412. B.A. Hills, Biophysical aspects of decompression, in P.B. Bennett and D.H. Elliott (Eds.), The Physiology and Medicine of

Diving and Compressed Air Work, Williams & Wilkins Co., Baltimore, 1975, pp. 366-391.

413. B.A. Hills, Zero-supersaturation approach to decompression optimization, in S.K. Hong (Ed.), International Symposium on Man in the Sea, Undersea Medical Society, Bethesda, 1975, pp. 179-189.

414. D.E. Yount, R.H. Strauss, E.L. Beckman and J.A. Moore, The physics of bubble formation: implications for improvement of decompression methods, in S.K. Hong (Ed.), International Symposium on Man in the Sea, Undersea Medical Society, Bethesda, 1975, pp. 167-178.

415. H. Yano, On improving ascent speed of decompression schedules, Ocean Age (May issue), (1978) 52-58.

416. Y. Mano, M. Shibayama, K. Ida, T. Miyamoto, H. Matzunaga, K. Ohgushi, A. Kashikura and H. Maeda, Comparative study of today's different decompression tables, Japn. J. Hyperbaric Med. 12 (1977) 14.

417. M.C. Marsh and F.P. Gorham, The gas disease in fishes, in Report of the U.S. Bureau of Fisheries 1904, U.S. Government Printing Office, Wash. DC, 1904, pp. 343-376.

418. M. Harris, W.E. Berg, D.M. Whitaker and V.C. Twitty, The relation of exercise to bubble formation in animals decompressed to sea level from high barometric pressures, J. Gen. Physiol. 28 (1945) 241-251.

419. P.M. McDonough and E.A. Hemmingsen, Bubble formation in crustaceans following decompression from hyperbaric gas exposures, J. Appl. Physiol. 56 (1984) 513-519.

420. P.M. McDonough and E.A. Hemmingsen, Bubble formation in crabs induced by limb motions after decompression, J. Appl. Physiol. 57 (1984) 117-122.

421. P.M. McDonough and E.A. Hemmingsen, A direct test for the survival of gaseous nuclei in vivo, Aviat. Space Environ. Med. 56 (1985) 54-56.

422. P.M. McDonough and E.A. Hemmingsen, Swimming movements initiate bubble formation in fish decompressed from elevated gas pressures, Comp. Biochem. Physiol. 81 (1985) 209-212.

423. J. How, A. Vijayan and T.M. Wong, Acute decompression sickness in compressed air workers exposed to pressures below 1 bar in the Singapore Mass Rapid Transit project, Singapore Med. J. 31 (1990) 104-110.

292

424.　B.B. Hemmingsen and E.A. Hemmingsen, Tolerance of bacteria to extreme gas supersaturations, Biochem. Biophys. Res. Commun. 85 (1978) 1379-1384.

425.　E.A. Hemmingsen, Cinephotomicrographic observations on intracellular bubble formation in *Tetrahymena*, J. Exp. Zool. 220 (1982) 43-48.

426.　E.A. Hemmingsen and B.B. Hemmingsen, Lack of intracellular bubble formation in microorganisms at very high gas supersaturations, J. Appl. Physiol. 47 (1979) 1270-1277.

427.　M.P. Spencer, D.C. Johanson and S.D. Campbell, Safe decompression with the Doppler ultrasonic blood bubble detector, in C.J. Lambertsen (Ed.), Underwater Physiology V, FASEB, Bethesda, 1976, pp. 311-325.

428.　A.V. Nebeker and J.R. Brett, Effects of air-supersaturated water on survival of Pacific salmon and steelhead smolts, Trans. Am. Fish. Soc. 105 (1976) 338-342.

429.　A. Evans and D.N. Walder, Significance of gas micronuclei in the aetiology of decompression sickness, Nature 222 (1969) 251-252.

430.　R.D. Vann, J. Grimstad and C.H. Nielsen, Evidence for gas nuclei in decompressed rats, Undersea Biomed. Res. 7 (1980) 107-112.

431.　E.N. Harvey, A.H. Whiteley, W.D. McElroy, D.C. Pease and D.K. Barnes, Bubble formation in animals, II, Gas nuclei and their distribution in blood and tissues, J. Cell. Comp. Physiol. 24 (1944) 23-34.

432.　B.G. D'Aoust and L.S. Smith, Bends in fish, Comp. Biochem. Physiol. 49A (1974) 311-321.

433.　E.N. Harvey, K.W. Cooper and A.H. Whiteley, Bubble formation from contact of surfaces, J. Am. Chem. Soc. 68 (1946) 2119-2120.

434.　A.T.J. Hayward, Tribonucleation of bubbles, Br. J. Appl. Phys. 18 (1967) 641-644.

435.　J. Campbell, The tribonucleation of bubbles, J. Phys. D Ser. 2 1 (1968) 1085-1088.

436.　K.G. Ikels, Production of gas bubbles in fluids by tribonucleation, J. Appl. Physiol. 28 (1970) 524-527.

437.　A.G. MacDonald, I. Gilchrist, K.T. Wann and S.E. Wilcock, The tolerance of animals to pressure, in R. Gilles (Ed.), Animals and Environmental Fitness, Pergamon, Oxford, 1980, pp.385-403.

438. P.K. Weathersby, L.D. Homer and E.T. Flynn, Homogeneous nucleation of gas bubbles in vivo, J. Appl. Physiol. 53 (1982) 940-946.

439. D.R. Gross, D.L. Miller and A.R. Williams, A search for ultrasonic cavitation within the mammalian cardiovascular system, Fed. Proc. 43 (1984) 297.

440. A. Shiina, K. Kondo, Y. Nakasone, M. Tsuchiya, T. Yaginuma and S. Hosoda, Contrast echocardiographic evaluation of changes in flow velocity, Circulation 63 (1981) 1408-1416.

441. R.S. Meltzer, B. Diebold, N.K. Valk, D. Blanchard, J.L. Guermonprez, C.T. Lancee, P. Peronneau and J. Roelandt, Correlation between velocity measurements from Doppler echocardiography and from M-mode contrast echocardiography, Br. Heart J. 49 (1983) 244-249.

442. R.A. Levine, L.E. Teichholz, M.E. Goldman, M.Y. Steinmetz, M. Baker and R.S. Meltzer, Microbubbles have intracardiac velocities similar to those of red blood cells, J. Am. Col. Cardiol. 3 (1984) 28-33.

443. S. Feinstein, K. Ong, Y. Fujibayashi, H. Staniloff, W. Zwehl, S. Meerbaum, E. Corday and P. Shah, Myocardial contrast echo disappearance rate is related to size and uniformity of sonicated agent microbubbles, Clin. Res. 32 (1984) 163A.

444. J. Ophir and K.J. Parker, Contrast agents in diagnostic ultrasound, Ultrasound Med. Biol. 15 (1989) 319-333.

445. B.B. Goldberg, J.B. Liu and F. Forsberg, Ultrasound contrast agents: a review, Ultrasound Med. Biol. 20 (1994) 319-333.

446. M.J.K. Blomley, J.C. Cooke, E.C. Unger, M.J. Monaghan and D.O. Cosgrove, Microbubble contrast agents: a new era in ultrasound, BMJ 322 (2001) 1222-1225.

447. S.I. Fox, Human Physiology, W.C. Brown, Dubuque, 1984, 715 pp.

448. A.J. Vander, J.H. Sherman and D.S. Luciano, Human Physiology: The Mechanisms of Body Function, 3rd ed., McGraw-Hill, New York, 1980, 724 pp.

449. G.J. Tortora, R.L. Evans and N.P. Anagnostakos, Principles of Human Physiology, Harper & Row, New York, 1982, 674 pp.

450. C. Tanford, The Hydrophobic Effect: Formation of Micelles and Biological Membranes, Wiley, New York, 1973, 200 pp.

451. N.A. Larichev, A.N. Gurov and V.B. Tolstoguzov, Protein-polysaccharide complexes at the interphase, 1, Characteristics of

294

decane/water emulsions stabilized by complexes of bovine serum albumin with dextran, Colloids Surfaces 6 (1983) 27-34.

452. A.N. Gurov, M.A. Mukhin, N.A. Larichev, N.V. Lozinskaya and V.B. Tolstoguzov, Emulsifying properties of proteins and polysaccharides, I, Methods of determination of emulsifying capacity and emulsion stability, Colloids Surfaces 6 (1983) 35-42.

453. J.H. Schulman and J.A. Friend, Light scattering investigation of the structure of transparent oil-water disperse systems, II, J. Colloid Sci. 4 (1949) 497-509.

454. J.H. Schulman, W. Stoeckenius and L.M. Prince, Mechanism of formation and structure of microemulsions by electron microscopy, J. Phys. Chem. 63 (1959) 1677-1680.

455. B.W. Barry, The control of oil-in-water emulsion consistency using mixed emulsifiers, J. Pharm. Pharmac. 21 (1969) 533-540.

456. B.W. Barry and G.M. Eccleston, Influence of gel networks in controlling consistency of o/w emulsions stabilised by mixed emulsifiers, J. Texture Stud. 4 (1973) 53-81.

457. S.S. Davis and A. Smith, The stability of hydrocarbon oil droplets at the surfactant/oil interface, Colloid Polym. Sci. 254 (1976) 82-98.

458. S. Fukushima, M. Takahashi and M. Yamaguchi, Effect of cetostearyl alcohol on stabilization of oil-in-water emulsion, J. Colloid Interface Sci. 57 (1976) 201-206.

459. M.J. Rosen, Surfactants and Interfacial Phenomena, Wiley, New York, 1978, 304 pp.

460. L.R. Angel, D.F. Evans and B.W. Ninham, Three-component ionic microemulsions, J. Phys. Chem. 87 (1983) 538-540.

461. E.W. Kaler, H.T. Davis and L.E. Scriven, Toward understanding microemulsion microstructure, II, J. Chem. Phys. 79 (1983) 5685-5692.

462. K. Larsson, Stability of emulsions formed by polar lipids, Prog. Chem. Fats other Lipids 16 (1978) 163-169.

463. A.W. Schwab, H.C. Nielsen, D.D. Brooks and E.H. Pryde, Triglyceride/aqueous ethanol/1-butanol microemulsions, J. Dispersion Sci. Technol. 4 (1983) 1-17.

464. B.J. Berne and R. Pecora, Dynamic Light Scattering, Wiley, New York 1976.

465. B.E. Dahneke (Ed.), Measurement of Suspended Particles by Quasi-Elastic Light Scattering, Wiley, New York, 1983, 570 pp.

466. A. Flamberg and R. Pecora, Dynamic light scattering study of micelles in a high ionic strength solution, J. Phys. Chem. 88 (1984) 3026-3033.

467. S.W. Provencher, Inverse problems in polymer characterization by direct analysis of polydispersity with photon correlation spectroscopy, Makromol. Chem. 180 (1979) 201-209.

468. S.W. Provencher, J. Hendrix, L. DeMayer and N. Paulussen, Direct determination of molecular-weight distributions of polystyrene in cyclohexane with photon correlation spectroscopy, J. Chem. Phys. 69 (1978) 4273-4276.

469. S.W. Provencher, European Molecular Biology Laboratory Technical Report EMBL-DA02, Heidelberg, 1980.

470. S.W. Provencher, A constrained regularization method for inverting data represented by linear algebraic or integral equations, Comput. Phys. Commun. 27 (1982) 213-227.

471. S. Bott, in B.E. Dahneke (Ed.), Measurement of Suspended Particles by Quasi-Elastic Light Scattering, Wiley, New York, 1983, p. 129.

472. S. Hayashi and S. Ikeda, Micelle size and shape of sodium dodecyl-sulfate in concentrated NaCl solutions, J. Phys. Chem. 84 (1980) 744-751.

473. S. Ikeda, S. Ozeki and M. Tsunoda, Micelle molecular weight of dodecyldimethylammonium chloride in aqueous solutions, and the transition of micelle shape in concentrated NaCl solutions, J. Colloid Interface Sci. 73 (1980) 27-37.

474. P.J. Missel, N.A. Mazer, G.B. Benedek, C.Y. Young and M.C. Carey, Thermodynamic analysis of the growth of sodium dodecyl-sulfate micelles, J. Phys. Chem. 84 (1980) 1044-1057.

475. S. Ikeda, S. Hayashi and T. Imae, Rodlike micelles of sodium dodecyl-sulfate in concentrated sodium-halide solutions, J. Phys. Chem. 85 (1981) 106-112.

476. G. Porte and J. Appell, Growth and size distributions of cetyl pyridinium bromide micelles in high ionic strength aqueous solutions, J. Phys. Chem. 85 (1981) 2511-2519.

477. S. Ozeki and S. Ikeda, The sphere-rod transition of micelles and the 2-step micellization of dodecyltrimethylammonium bromide in aqueous NaBr solutions, J. Colloid Interface Sci. 87 (1982) 424-435.

478. S. Ozeki and S. Ikeda, The sphere-rod transition of micelles of dodecyldimethylammonium bromide in aqueous NaBr solutions,

and the effects of counterion binding on the micelle size, shape and structure, Colloid Polym. Sci. 262 (1984) 409-417.

479. T. Imae and S. Ikeda, Rodlike micelles of dimethyloleylamine oxide in aqueous NaCl solutions, and their flexibility and size distribution, Colloid Polym. Sci. 262 (1984) 497-506.

480. U. Kaatze, S.C. Muller and H. Eible, Monoalkyl phosphodiesters synthesis and dielectric relaxation of solutions, Chem. Phys. Lipids 27 (1980) 263-280.

481. D.F. Nicoli, J.G. Elias and D. Eden, Transient electric birefringence study of CTAB micelles: implications for rodlike growth, J. Phys. Chem. 85 (1981) 2866-2869.

482. S. Ozeki and S. Ikeda, The viscosity behavior of aqueous NaCl solutions of dodecyldimethylammonium chloride and the flexibility of its rod-like micelle, J. Colloid Interface Sci. 77 (1980) 219-231.

483. R. Nagarajan, Are large micelles rigid or flexible: a reinterpretation of viscosity data for micellar solutions, J. Colloid Interface Sci. 90 (1982) 477-486.

484. Er. Gulari, Es. Gulari, Y. Sunashima and B. Chu, Polymer diffusion in a dilute theta solution, 1, Polystyrene in cyclohexane, Polymer 20 (1979) 347-355.

485. D.E. Koppel, Analysis of macromolecular polydispersity in intensity correlation spectroscopy: method of cumulants, J. Chem. Phys. 57 (1972) 4814-4820.

486. R.R. Balmbra, J.S. Clunie, J.M. Corkill and J.F. Goodman, Effect of temperature on the micelle size of a homogeneous nonionic detergent, Trans. Faraday Soc. 58 (1962) 1661-1667.

487. R.R. Balmbra, J.S. Clunie, J.M. Corkill and J.F. Goodman, Variations in the micelle size of nonionic detergents, Trans. Faraday Soc. 60 (1964) 979-985.

488. P.H. Elworthy and C.B. Macfarlane, Chemistry of nonionic detergents, V, Micellar structures of a series of synthetic nonionic detergents, J. Chem. Soc. (1963) 907-914.

489. P.H. Elworthy and C. McDonald, Chemistry of nonionic detergents, VII, Variations of the micellar structure of some synthetic nonionic detergents with temperature, Kolloid Z.Z. Polym. 195 (1964) 16-23

490. K.W. Herrmann, J.G. Bruchmiller and W.L. Courchene, Micellar properties and critical opalescence of dimethylalkylphosphine oxide solutions, J. Phys. Chem. 70 (1966) 2909-2918.

491.   J.M. Corkill, J.F. Goodman and T. Walker, Light scattering from aqueous solutions of octylsulfinylalkanols, Trans. Faraday Soc. 63 (1967) 759-767.

492.   R.H. Ottewill, C.C. Storer and T. Walker, Ultracentrifugal study of a nonionic surface-active agent in D_2O, Trans. Faraday Soc. 63 (1967) 2796-2802.

493.   D. Attwood, Light-scattering study of the effect of temperature on the micellar size and shape of a nonionic detergent in aqueous solution, J. Phys. Chem. 72 (1968) 339-345.

494.   G.L.K. Hoh, D.O. Barlow, A.F. Chadwick, D.B. Lake and S.R. Sheeran, Hydrogen peroxide oxidation of tertiary amines, J. Am. Oil Chem. Soc. 40 (1963) 268-271.

495.   S. Ikeda, M. Tsunoda and H. Maeda, Application of Gibbs adsorption isotherm to aqueous solutions of a nonionic cationic surfactant, J. Colloid Interface Sci. 67 (1978) 336-348.

496.   K.W. Herrmann, Nonionic-cationic micellar properties of di-methyldodecylamine oxide, J. Phys. Chem. 66 (1962) 295-300.

497.   J.M. Corkill and K.W. Herrmann, Solution structure in concentrated nonionic surfactant systems, J. Phys. Chem. 67 (1963) 934-937.

498.   S. Ikeda, M. Tsunoda and H. Maeda, Effects of ionization on micelle size of dimethyldodecylamine oxide, J. Colloid Interface Sci. 70 (1979) 448-455.

499.   W.L. Courchene, Micellar properties from hydrodynamic data, J. Phys. Chem. 68 (1964) 1870-1874.

500.   T. Imae and S. Ikeda, Intermicellar correlation in light-scattering from dilute micellar solutions of dimethyloleylamine oxide, J. Colloid Interface Sci. 98 (1984) 363-372.

501.   P.L. Bolden, J.C. Hoskins and A.D. King, Jr., The solubility of gases in solutions containing sodium alkylsulfates of various chain lengths, J. Colloid Interface Sci. 91 (1983) 454-463.

502.   P.A. Winsor, Solvent Properties of Amphiphilic Compounds, Butterworths, London, 1954.

503.   M.E.L. McBain and E. Hutchinson, Solubilization and Related Phenomena, Academic Press, New York, 1955.

504.   P.H. Elworthy, A.T. Florence and C.B. Macfarlane, Solubilization by Surface Active Agents, Chapman & Hall, London, 1968.

505.   I.B.C. Matheson and A.D. King, Jr., Solubility of gases in micellar solutions, J. Colloid Interface Sci. 66 (1978) 464-469.

298

506.   D.W. Ownby and A.D. King, Jr., The solubility of propane in micellar solutions of sodium octyl sulfate and sodium dodecyl sulfate at 25, 35, and 45°C, J. Colloid Interface Sci. 101 (1984) 271-276.

507.   J.W. McBain and J.J. O'Conner, A simple proof of the thermodynamic stability of materials taken up by solutions containing solubilizers such as soap, J. Am. Chem. Soc. 62 (1940) 2855-2859.

508.   J.W. McBain and J.J. O'Conner, Effect of potassium oleate on the solubility of hydrocarbon vapors in water, J. Am. Chem. Soc. 63 (1941) 875-877.

509.   J.W. McBain and A.M. Soldate, The solubility of propylene vapor in water as affected by typical detergents, J. Am. Chem. Soc. 64 (1942) 1556-1557.

510.   S. Ross and J.B. Hudson, Henry's-law constants of butadiene in aqueous solutions of a cationic surfactant, J. Colloid Sci. 12 (1957) 523-525.

511.   A. Wishnia, The hydrophobic contribution to micelle formation: the solubility of ethane, propane, butane, and pentane in sodium dodecyl sulfate solution, J. Phys. Chem. 67 (1963) 2079-2082.

512.   L.J. Winters and E. Grunwald, Organic reactions occurring in or on micelles, III, Reaction of methyl bromide with cyanide ion, J. Am. Chem. Soc. 87 (1965) 4608-4611.

513.   K.W. Miller, L. Hammond and E.G. Porter, The solubility of hydrocarbon gases in lipid bilayers, Chem. Phys. Lipids 20 (1977) 229-241.

514.   J.C. Hoskins and A.D. King, Jr., The effect of n-pentanol on the solubility of ethane in micellar solutions of sodium dodecyl sulfate, J. Colloid Interface Sci. 82 (1981) 260-263.

515.   J.C. Hoskins and A.D. King, Jr., The effect of sodium chloride on the solubility of ethane in micellar solutions of sodium dodecyl sulfate, J. Colloid Interface Sci. 82 (1981) 264-267.

516.   S.D. Christian, E.E. Tucker and E.H. Lane, Precise vapor-pressure measurements of the solubilization of cyclohexane by sodium octyl sulfate and sodium octyl sulfate micelles, J. Colloid Interface Sci. 84 (1981) 423-432.

517.   L. DellaGuardia and A.D. King, Jr., Contact charge-transfer spectra arising from anisole and molecular oxygen cosolubilized in micelles of sodium dodecyl sulfate at elevated pressures, J. Colloid Interface Sci. 88 (1982) 8-16.

518. J. Wilf and A. Ben-Naim, Solubilization of paraffin gases in aqueous solutions of sodium octanoate, J. Solution Chem. 12 (1983) 671-683.

519. K.A. Zachariasse, B. Kozankiewicz and W. Kuhnle, Micellar structure and water penetration studied by NMR and optical spectroscopy, in K.L. Mittal and B. Lindman (Eds.), Surfactants in Solution, Plenum, New York, 1984, pp.565-584.

520. D.G. Whitten, J.B.S. Bonilha, K.S. Schanze and J.R. Winkle, Solubilization and water penetration into micelles and other organized assemblies as indicated by photochemical studies, in K.L. Mittal and B. Lindman (Eds.), Surfactants in Solution, Plenum, New York, 1984, pp. 585-598.

521. B.A. Hills, Air embolism: fission of microbubbles upon collision in plasma, Clin. Sci. Mol.. Med. 46 (1974) 629-634.

522. B.W. Muller and R.H. Muller, Particle size analysis of latex suspensions and microemulsions by photon correlation spectroscopy, J. Pharm. Sci. 73 (1984) 915-918.

523. B.W. Muller and R.H. Muller, Particle size distributions and particle size alterations in microemulsions, J. Pharm. Sci. 73 (1984) 919-922.

524. J.S. D'Arrigo, Contrast-assisted tumor detection, Drug News Perspectives 4 (1991) 164-167.

525. J.S. D'Arrigo, R.H. Simon and S.Y. Ho, Lipid-coated uniform microbubbles for earlier sonographic detection of brain tumors, J. Neuroimagimg 1 (1991) 134-139.

526. R.H. Simon, S.Y. Ho, C.R. Perkins and J.S. D'Arrigo, A quantitative assessment of tumor enhancement by ultrastable lipid-coated microbubbles as a contrast agent, Invest. Radiol. 27 (1992) 29-34.

527. J.S. D'Arrigo and T. Imae, Physical characteristics of ultrastable lipid-coated microbubbles, J. Colloid Interface Sci.149 (1992) 592-595.

528. J.S. D'Arrigo, S.Y. Ho and R.H. Simon, Detection of experimental rat liver tumors by contrast-assisted ultrasonography, Invest. Radiol. 28 (1993) 218-222.

529. W. Huang, J.C. Grecula, T.M. Button, D.P. Harrington, M.A. Davis, J.S. D'Arrigo, B.H. Laster and C.S. Springer, Use of lipid-coated microbubbles (LCMs) for susceptibility-based MRI contrast in brain tumors, Proceedings of the 12th Annual Meeting of the Society of Magnetic Resonance in Medicine, New York,

1993.

530. R.H. Simon, S.Y. Ho, D.F. Uphoff, S.C. Lange and J.S. D'Arrigo, Applications of lipid-coated microbubble ultrasonic contrast to tumor therapy, Ultrasound Med. Biol. 19 (1993) 123-125.

531. E. Barbarese, S.Y. Ho, J.S. D'Arrigo and R.H. Simon, Internalization of microbubbles by tumor cells in vivo and in vitro, J. Neuro-Oncology 26 (1995) 25-34.

532. S.Y. Ho, E. Barbarese, J.S. D'Arrigo, C. Smith and R.H. Simon, Evaluation of lipid-coated microbubbles as a delivery vehicle for Taxol in tumor therapy, Neurosurgery 40 (1997) 1260-1268.

533. S.Y. Ho, X.G. Li, A. Wakefield, E. Barbarese, J.S. D'Arrigo and R.H. Simon, The affinity of lipid-coated microbubbles for maturing brain injury sites, Brain Res. Bull. 43 (1997) 543-549.

534. A.E. Wakefield, S.Y. Ho, X.G. Li, J.S. D'Arrigo and R.H. Simon, The use of lipid-coated microbubbles as a delivery agent for 7β-hydroxycholesterol to a radiofrequency lesion in the rat brain, Neurosurgery 42 (1998) 592-598.

535. I.U. Kureshi, S.Y. Ho, H.C. Onyiuke, A.E. Wakefield, J.S. D'Arrigo and R.H. Simon, The affinity of lipid-coated microbubbles to maturing spinal cord injury sites, Neurosurgery 44 (1999) 1047-1053.

536. R.H. Simon, S.Y. Ho, J.S. D'Arrigo, A. Wakefield and S.G. Hamilton, Lipid-coated ultrastable microbubbles as a contrast agent in neurosonography, Invest. Radiol. 25 (1990) 1300-1304.

537. H.F. Stewart and M.E. Stratmeyer (editors), An Overview of Ultrasound: Theory, Measurement, Medical Applications and Biological Effects, HHS Publication FDA 82-8190, Rockville MD, 1982, 134 pp.

538. J.S. D'Arrigo, Method for the production of medical-grade lipid-coated microbubbles, paramagnetic labeling of such microbubbles and therapeutic uses of microbubbles, United States Patent No. 5,215,680 (issued 1993).

539. ibid. (2nd edition, revised), Australia Patent No. 657480 (issued 1995).

540. ibid. (3rd edition, revised), United Kingdom Patent No. 0467031 (issued 1997).

541. ibid. (3rd edition, revised), Germany Patent No. 69127032.5 (issued 1998).

542. ibid. (3rd edition, revised), France Patent No. 0467031 (issued

1997).

543. ibid. (3rd edition, revised), Italy Patent No. 0467031 (issued 1997).

544. J.S. D'Arrigo, Surfactant mixtures, stable gas-in-liquid emulsions, and methods for the production of such emulsions from said mixtures, United States Patent No. 4,684,479 (issued 1987).

545. ibid. (2nd edition, revised), Canada Patent No. 1,267,055 (issued 1990).

546. ibid. (3rd edition, revised), Japan Patent No. 1,815,442 (issued 1994).

547. F. Muller-Landau and D.A. Cadenhead, Molecular packing in steroid-lecithin monolayers, Part I, Pure films of cholesterol, 3-doxyl-cholestane, 3-doxyl-17-hydroxyl-androstane, tetradecanoic acid and dipalmitoylphosphatidylcholine, Chem. Phys. Lipids 25 (1979) 299-314.

548. F. Muller-Landau and D.A. Cadenhead, Molecular packing in steroid-lecithin monolayers, Part II, Mixed films of cholesterol with dipalmitoylphosphatidylcholine, and tetradeanoic acid, Chem. Phys. Lipids 25 (1979) 315-328.

549. F. Muller-Landau and D.A. Cadenhead, Molecular packing in steroid-lecithin monolayers, Part III, Mixed films of 3-doxyl-cholestane and 3-doxyl-17-hydroxyl-androstane with dipal-mitoyl-phosphatidylcholine, Chem. Phys. Lipids 25 (1979) 329-343.

550. M. Sugahara, M. Uragami, X. Yan and S.L. Regen, The structural role of cholesterol in biological membranes, J. Am. Chem. Soc. 123 (2001) 7939-7940.

551. M. Sugahara, M. Uragami, N. Tokutake and S.L. Regen, The importance of acyl chain placement on phospholipid mixing in the physiologically relevant fluid phase, J. Am. Chem. Soc. 123 (2001) 2697-2698.

552. M. Kako and S. Kondo, The stability of soybean oil-water emulsions containing mono- and diglcerides, J. Colloid Interfact Sci. 69 (1979) 163-169.

553. A. Rahman and P. Sherman, Interaction of milk proteins with monoglycerides and diglycerides, Colloid Polym. Sci. 260 (1982) 1035-1041.

554. G. Doxastakis and P. Serman, The interaction of sodium caseinate with monglyceride and diglyceride at the oil-water

302

interface and its effect on interfacial rheological properties, Colloid Polym. Sci. 264 (1986) 254-259.

555. D.F. Evans and B.W. Ninham, Molecular forces in the self-organization of amphiphiles, J. Phys. Chem. 90 (1986) 226-234.

556. R.H. Yoon and J.L. Yorden, Zeta-potential measurements on microbubbles generated using various surfactants, J. Colloid Interface Sci. 113 (1986) 430-438.

557. H. Steinbach and C. Sucker, Structures of association in surface films, Adv. Colloid Interface Sci. 14 (1980) 43-65.

558. J. Clifford and B.A. Pethica, Properties of micellar solutions, Part 2, N.m.r. chemical shift of water protons in solutions of sodium alkyl sulphates, Trans. Faraday doc. 60 (1964) 1483-1490.

559. L. Benjamin, Partial molal volume changes during micellization and solution of nonionic surfactants and perfluorocarboxulates using a magnetic density balance, J. Phys. Chem. 70 (1966) 3790-3797.

560. J.M. Corkill, J.F. Goodman and T. Walker, Partial molar volumes of surface-active agents in aqueous solution, Trans. Faraday Soc. 63 (1967) 786-772.

561. T. Walker, The influence of surface active agents on the structure of water, J.Colloid Interface Sci. 45 (1973) 372-377.

562. L.F. Scatena, M.G. Brown and G.L. Richmond, Water at hydrophobic surfaces: weak hydrogen bonding and strong orientation effects, Science 292 (2001) 908-912.

563. V.T. Mallory, Acute intravenous (i.v.) toxicity study in rabbits: Filmix, Report No. PH 422-CV-001-90, Pharmakon Research International Inc., Waverly PA, 1990.

564. V.T. Mallory, Acute intravenous (i.v.) toxicity study in dogs: Filmix, Report No. PH 444-CV-001-90, Pharmakon Research International Inc., Waverly PA, 1990.

565. J.C.E. Underwood and I. Carr, The ultrastructure and permeability characteristics of the blood vessels of a transplantable rat sarcoma, J. Pathol. 107 (1972) 157-165.

566. D.M. Long, Capillary ultrastructure and the blood-brain barrier in human malignant brain tumors, J. Neurosurg. 32 (1970) 127-144.

567. J.D. Ward, M.G. Hadfield, D.P. Becker and E.T. Lovings, Endothelial fenestrations and other vascular alterations in primary melanoma of the central nervous system, Cancer 34 (1974) 1982-1991.

568. P.A. Stewart, K. Hayakawa, E. Hayakawa, C.L. Farrell and R.F.

Del Maestro, A quantitative study of blood-brain barrier permeability ultrastructure in a new rat glioma model, Acta Neuropath. 67 (1985) 96-102.

569. Schultz, Methods for fats and lipids, in G.L. Lee (Ed.), Armed Forces Institute of Pathology Manual of Histologic Staining Methods, 1968, p. 140.

570. M.A. Davis, S.A. Scatamacchia, J.S. D'Arrigo, R.H. Simon and R.H. Wrigley, Lipid-coated microbubbles (Filmix®) as a sonographic contrast agent, Med. Physics 19 (1992) 1138.

571. R.D. Lillie and L.L. Ashburn, Supersaturated solutions of fat stains in dilute isopropanol for demonstration of acute fatty degenerations not shown by Herxheimer Technic, Arch. Pathol. 36 (1943) 432.

572. F. Forsberg, W.T. Shi and B.B. Goldberg, Subharmonic imaging of contrast agents, Ultrasonics 38 (2000) 93-98.

573. P.J. Frinking, A. Bouakaz, J. Kirkhorn, F.J. Ten Cate and N. de Jong, Ultrasound contrast imaging: current and new potential methods, Ultrasound Med. Biol. 26 (2000) 965-975.

574. P.D. Krishna, P.M. Shankar and V.L. Newhouse, Subharmonic generation from ultrasonic contrast agents, Phys. Med. Biol. 44 (1999) 681-694.

575. K.I. Morton, G.R. Ter Haar, I.J. Stratford and C.R. Hill, Subharmonic emission as an indicator of ultrasonically-induced biological damage, Ultrasound Med. Biol. 9 (1983) 629-633.

576. R.J. Jeffers, R.Q Feng, J.B. Fowlkes, J.W. Hunt, D. Kessel and C.A. Cain, Dimethylformamide as an enhancer of cavitation-induced cell lysis in vitro, J. Acoust. Soc. Am. 97 (1995) 669-676.

577. M.S. Albert, W. Huang, J.H. Lee, J.A. Balschi and C.S. Springer, Aqueous shift reagents for high-resolution cation NMR, VI, Titration curves for in vivo ^{23}Na and ^{1}H$_2$O MRS obtained from rat blood, NMR Biomed. 6 (1993) 7.

578. G.F. Whalen, Lipid coated microbubble (LCM)-facilitated ultrasonic treatment of liver tumors, Clinical Protocol No. 99-235 filed with Office of Clinical Research, Institutional Review Board (IRB), Univ. of Connecticut Health Ctr. (IRB Approval by full Board for first phase granted 5-13-99) in conjunction with Critical Technologies Funding Competition (G.F. Whalen and J.S. D'Arrigo, Co-P.I.'s), 1999 (limited distribution reports).

579. R. Yang, C.R. Reilly, F.J. Rescorla, P.R. Faught, N.T. Sanghvi,

304

F.J. Fry, T.D. Franklin, L. Lumeng and J.L. Grosfeld, High-intensity focused ultrasound in the treatment of experimental liver cancer, Arch. Surg. 126 (1991) 1002-1010.

580. V.K. Rustgi (A.M. DiBisceglie, moderator), Epidemiology of hepatocellular carcinoma: hepatocellular carcinoma, Ann. Intern. Med. 108 (1988) 390-401.

581. J.S. Stehlin, Jr., P.D. Ipolyi, P.J. Greeff, C.J. McGaff,Jr., B.R. Davis and L. McNary, Treatment of cancer of the liver: twenty years' experience with infusion and resection of 414 patients, Ann. Surg. 208 (1988) 23-25.

582. J.W. Pickren, Liver metastasis: analysis of autopsies, in L. Weiss and H.A. Gilbert (Eds.), Liver Metastasis, G.K. Hall & Co., Boston, 1981, pp. 2-18.

583. M.G. Honig and R.I. Hume, Fluorescent carbocyanine dyes allow living neurons of identified origin to be studied in long-term cultures, J. Cell Biol. 103 (1986) 171-187.

584. E. Barbarese, H. Soares, S. Yang and R.B. Clark, Comparison of CNS homing pattern among murine T cell lines responsive to myelin basic protein, J. Neuroimmunol. 39 (1992) 151-162.

585. F. Morgan, E. Barbarese and J. Carson, Visualizing cells in three dimensions using confocal microscopy, image reconstruction and isosurface rendering: applications to glial cells in mouse CNS, Scanning Micros. 6 (1992) 345-358.

586. K. Ainger, D. Avossa, F. Morgan, S.J. Hill, C. Barry, E. Barbarese and J.H. Carson, Transport and localization of exogenous myelin basic protein mRNA microinjected into oligodendrocytes, J. Cell Biol. 123 (1993) 431-441.

587. D.L. Farkas, M.D. Wei, P. Febbroiello, J.H. Carson and L.M. Loew, Simultaneous imaging of cell and mitochondrial membrane potentials, Biophys. J. 56 (1989) 1053-1069.

588. L. Acarin, J.M. Vela, B. Gonzalez and B. Castellano, Demonstration of poly-N-acetyl lactosamine residues in ameboid and ramified microglial cells in rat brain by tomato lectin binding, J. Histochem. Cytochem. 42 (1994) 1033-1041.

589. R.D. Fross, P.C. Warnke and D.R. Groothuis, Blood flow and blood-to-tissue transport in 9L gliosarcomas: the role of the brain tumor model in drug delivery research, J. Neuro-Oncol. 11 (1991) 185-197.

590. G. Dapergolas, Penetration of target areas in the rat by liposome-associated bleomycin, glucose oxidase and insulin, FEBS Lett. 63

(1976) 235-240.

591. G. Gregoriadis, J. Senior, B. Wolff and C. Kirby, Targeting of liposomes to accessible cells in vivo, Ann. N.Y. Acad. Sci. 446 (1985) 319-340.

592. G. Gregoriadis, Targeting of drugs: implications in medicine, Lancet 2 (1981) 241-247.

593. Y. Hashimoto, M. Sugaware, T. Masuko and H. Hojo, Anti-tumor effect of actinomycin D entrapped in liposomes bearing subunits of tumor-specific monoclonal immunoglobulin M antibody, Cancer Res. 43 (1983) 5328-5334.

594. R.T. Proffitt, L.E. Williams, C.A. Presant, G.W. Tin, J.A. Uliana, R.C. Gamble and J.D. Baldeschwieler, Liposomal blockade of the reticuloendothelial system: improved tumor imaging with small unilamellar vesicles, Science 220 (1983) 502-505.

595. G. Griffiths, B. Hoflack, K. Simons, I. Mellman and S. Kornfeld, The mannose 6-phosphate receptor and the biogenesis of lysosomes, Cell 52 (1988) 329-341.

596. W.P. McGuire, E.K. Rowinsky, N.B. Rosenshein, F.C. Grumbine, D.S. Ettinger, D.K. Armstrong and R.C. Donehower, Taxol: a unique antineoplastic agent with significant activity in advanced ovarian epithelial neoplasms, Ann. Intern. Med. 111 (1989) 273-279.

597. National Cancer Institute, Clinical Brochure: Taxol (NSC 125973), Division of Cancer Treatment-NCI, Bethesda MD, 1983.

598. W.C. Rose, Taxol: a review of its preclinical in vivo antitumor activity, Anticancer Drugs 3 (1992) 311-321.

599. E.K. Rowinsky, L.A. Cazenave and R.C. Donehower, Taxol: a novel investigational antimicrotubule agent, J. Natl. Cancer Inst. 82 (1990) 1247-1259.

600. M. De Brabander, G. Geuens, R. Nuydens, R. Willebrords and J. De Mey, Taxol induces the assembly of free microtubules in living cells and blocks the organizing capacity of the centrosome and kinetochores, Proc. Natl. Acad. Sci. USA 78 (1981) 5608-5612.

601. M.A. Jordan, R.J. Toso, D. Thrower and L. Wilson, Mechanism of mitotic block and inhibition of cell proliferation by Taxol at low concentrations, Proc. Natl. Acad. Sci. USA 90 (1993) 9552-9556.

602. W. Lorenz, H.J. Reimann and A. Schmal, Histamine release in

306

dogs by cremophor-EL and its derivatives: oxyethylated oleic acid is the most effective constituent, Agents Actions 7 (1977) 63-67.

603.   R.B. Weiss, T.C. Donehower, P.H. Wiernik, T. Ohnuma, R.J. Gralla, D.I. Trump, J.R. Baker, Jr., D.A. Van Echo, D.D. Von Hoff and B. Leyland-Jones, Hypersensitivity reactions from Taxol, J. Clin. Oncol. 8 (1990) 1263-1268.

604.   J. Pellegrino, A.S. Pellegrino and A.J. Cushman, A Stereotaxic Atlas of the Rat Brain, 2nd ed., Plenum Press, New York, 1979.

605.   A. Sharma, E. Mayhew and R.M. Straubinger, Antitumor effect of Taxol-containing liposomes in a Taxol-resistant murine tumor model, Cancer Res. 53 (1993) 5877-5881.

606.   H. Maeda, L.W. Seymour and Y. Miyamoto, Conjugates of anticancer agents and polymers: advantages of macromolecular therapeutics in vivo, Bioconjugate Chem. 3 (1992) 351-362.

607.   D. Bochelen, F. Eclancher, A. Kupferberg, A. Privat and M. Mersel, 7β-hydroxycholesterol and 7β-hydroxycholesteryl-3-esters reduce the extent of reactive gliosis caused by an electrolytic lesion in rat brain, Neuroscience 51 (1992) 827-834.

608.   D. Bochelen, M. Mersel, P. Behr and P. Lutz, Effects of oxysterol treatment on cholesterol biosynthesis and reactive astrocyte proliferation in injured rat brain cortex, J. Neurochem. 65 (1995) 2194-2200.

609.   P.H.L. Hwang, Inhibitors of protein and RNA synthesis block the cytotoxic effects of oxygenated sterols, Biochim. Biophys. Acta 1136 (1992) 5-11.

610.   A. Kupferberg, G. Teller and P. Behr, Metabolism of 7β-hydroxycholesterol in astrocyte primary cultures and derived spontaneously transformed cell lines: correlation between the esterification on C-3-OH by naturally occurring fatty acids and cytotoxicity, Biochim. Biophys. Acta 1046 (1990) 106-109.

611.   A. Kupferberg, G. Teller, P. Behr, C. Leray, P.F. Urban and G. Vincendon, Effect of 7β-hydroxycholesterol on astrocyte primary cultures and derived spontaneously transformed cell lines: cytotoxicity and metabolism, Biochim. Biophys. Acta 1013 (1989) 231-238.

612.   R.A. Firestone, Selective delivery of cytotoxic compounds to cells by the LDL pathway, J. Med. Chem. 27 (1984) 1037-1043.

613.   S. Vitols, Uptake of low-density lipoprotein by malignant cells: possible therapeutic applications, Rev. Cancer Cells 3 (1991)

488-495.

614. A. Menrad and F.A. Anderer, Expression of LDL receptor on tumor cells induced by growth factors, Anticancer Res. 11 (1991) 385-390.

615. P.J. Pussinen, B. Karten, A. Wintersperger, H. Reicher, M. McLean, E. Malle and W. Sattler, The human breast carcinoma cell line HBL-100 acquires exogenous cholesterol from high-density lipoprotein via CLA-1 (CD-36 and LIMPII analogous 1)-mediated selective cholesteryl ester uptake, Biochem. J. 349 (2000) 559-566.

616. J.L. Goldstein, R.G.W. Anderson and M.S. Brown, Coated pits, coated vesicles, and receptor mediated endocytosis, Nature 279 (1979) 679-684.

617. J.L. Goldstein and M.S. Brown, The low-density lipoprotein pathway and its relation to atherosclerosis, Ann. Rev. Biochem. 46 (1977) 897-930.

618. J.L. Goldstein and M.S. Brown, A receptor-mediated pathway for cholesterol homeostasis, Science 232 (1986) 34-47.

619. S. Vitols, C. Peterson, O. Larsson, P. Holm and B. Aberg, Elevated uptake of low density lipoproteins by human lung cancer tissue in vivo, Cancer Res. 52 (1992) 6244-6247.

620. S.G. Young, Recent progress in understanding apolipoprotein B, Circulation 82 (1990) 1574-1594.

621. S.O. Olofsson, G. Bjursell, K. Bostrom, P. Carlsson, J. Elovson, A.A. Protter, M.A. Reuben and G. Bondjers, Apolipoprotein B: structure, biosynthesis and role in the lipoprotein assembly process, Atherosclerosis 68 (1987) 1-17.

622. S.A. Hynds, J. Welsh, J.M. Stewart, A. Jack, M. Soukop, C.S. McArdle, K.C. Calman, C.P. Packard and J. Sheperd, Low-density lipoprotein metabolism in mice with soft tissue tumors, Biochem. Biophys. Acta 795 (1984) 589-595.

623. G. Norata, G. Canti, L. Ricci, A. Nicolin, E. Trezzi and A.L. Catapano, In vivo assimilation of low density lipoprotein by a fibrosarcoma tumor line in mice, Cancer Lett. 25 (1984) 203-208.

624. Y.K. Ho, G.S. Smith, M.S. Brown and J.L. Goldstein, Low-density lipoprotein (LDL) receptor activity in human acute myelogenous leukemia cells, Blood 52 (1978) 1099-1114.

625. P. Lombardi, G. Norata, F.M. Maggi, G. Canti, P. Franco, A. Nicolin and A.L. Catapano, Assimilation of LDL by experimental tumours in mice, Biochim. Biophys. Acta 1003

(1989) 301-306.

626. S. Vitols, G. Gahrton, A. Ost and C. Peterson, Elevated low density lipoprotein receptor activity in leukemic cells with monocytic differentiation, Blood 63 (1984) 1186-1193.

627. M.J. Rudling, B. Angelin, C.O. Peterson and V.P. Collins, Low density lipoprotein receptor activity in human intracranial tumors and its relation to the cholesterol requirement, Cancer Res. 50 (1990) 483-487.

628. S. Vitols, G. Gahrton, M. Bjorkholm and C. Peterson, Hypocholesterolemia in malignancy due to elevated LDL receptor activity in tumour cells: evidence from studies of leukemic patients, Lancet 2 (1985) 1150-1154.

629. S. Vitols, G. Gahrton and C. Peterson, Significance of the low-density lipoprotein receptor pathway for the in vitro accumulation of AD-32 incorporated into LDL in normal and leukemic white blood cells, Cancer Treat. Rep. 68 (1984) 515-520.

630. M.J. Rudling, V.P. Collins and C.O. Peterson, Delivery of aclacinomycin A to human glioma cells in vitro by the low density lipoprotein pathway, Cancer Res. 43 (1983) 4600-4605.

631. G. Martin, A. Pilon, C. Albert, M. Valle, D.W. Hum, J.C. Fruehart, J. Najib, V. Clavey and B. Staels, Comparison of expression and regulation of the high-density lipoprotein receptor SR-BI and the low-density lipoprotein receptor in human adrenocortical carcinoma NCI-H295 cells, Eur. J. Biochem. 261 (1999) 481-491.

632. S.T. Mosley, J.L. Goldstein, M.S. Brown, J.R. Falck and R.G.W. Anderson, Targeted killing of cultured cells by receptor-dependent photosensitization, Proc. Natl. Acad. Sci. USA 78 (1981) 5717-5721.

633. S.G. Vitols, M. Masquelier and C.O. Peterson, Selective uptake of a toxic lipophilic anthracycline derivative by the low-density lipoprotein receptor pathway in cultured fibroblasts, J. Med. Chem. 28 (1985) 451-454.

634. R. Lindquist, S. Vitols, A. Ost, G. Gahrton and C. Peterson, Low-density lipoprotein receptor activity in human leukemic cells: relation to chromosome aberrations, Acta Med. Scand. 217 (1985) 553-558.

635. A.J. Versluis, E.T. Rump, P.C. Rensen, T.J. van Berkel and M.K. Bijsterbosch, Stable incorporation of a lipophilic daunorubicin prodrug into apolipoprotein E-exposing liposomes induces uptake

of prodrug via low-density lipoprotein receptor in vivo, J. Pharmacol. Exp. Ther. 289 (1999) 1-7.

636. P.C. Rensen, R.L. de Vrueh, J. Kuiper, M.K. Bijsterbosch, E.A. Biessen and T.J. van Berkel, Recombinant lipoproteins: lipoprotein-like lipid particles for drug targeting, Adv. Drug Deliv. Rev. 47 (2001) 251-276.

637. L. Maletinska, E.A. Blakely, K.A. Bjornstad, D.F. Deen, L.J. Knoff and T.M. Forte, Human glioblastoma cell lines: levels of low-density lipoprotein receptor and low-density lipoprotein receptor-related protein, Cancer Res. 60 (2000) 2300-2303.

638. A.J. Versluis, P.C. Rensen, E.T. Rump, T.J. Van Berkel and M.K. Bijsterbosch, Low-density lipoprotein receptor-mediated delivery of a lipophilic daunorubicin derivative to B16 tumours in mice using apolipoprotein E-enriched liposomes, Br. J. Cancer 78 (1998) 1607-1614.

639. A.J. Versluis, E.T. Rump, P.C. Rensen, T.J. Van Berkel and M.K. Bijsterbosch, Synthesis of a lipophilic daunorubicin derivative and its incorporation into lipidic carriers developed for LDL receptor-mediated tumor therapy, Pharm Res. 15 (1998) 531-537.

640. E.T. Rump, R.L. de Vrueh, L.A. Sliedregt, E.A. Biessen, T.J. van Berkel and M.K. Bijsterbosch, Preparation of conjugates of oligodeoxynucleotides and lipid structures and their interaction with low-density lipoprotein, Bioconjug. Chem. 9 (1998) 341-349.

641. P.C. Rensen, R.M. Schiffelers, A.J. Versluis, M.K. Bijsterbosch, M.E. Van Kuijk-Meuwissen and T.J. Van Berkel, Human recombinant apolipoprotein E-enriched liposomes can mimic low-density lipoproteins as carriers for the site-specific delivery of antitumor agents, Mol. Pharmacol. 52 (1997) 445-455.

642. D. Gal, M. Ohashi, P.C. MacDonald, H.J. Buchsbaum and E.R. Simpson, Low-density lipoprotein as a potential vehicle for chemotherapeutic agents and radionucleotides in the management of gynecologic neoplasms, Am. J. Obstet. Gynecol. 139 (1981) 877-885.

643. R.G.W. Anderson, M.S. Brown and J.L. Goldstein, Role of the coated endocytic vesicle in the uptake of receptor-bound low density lipoprotein in human fibroblasts, Cell 10 (1977) 351-364.

644. J.L. Goldstein, S.K. Basu, G.Y. Brunschede and M.S. Brown, Release of low density lipoprotein from its cell surface receptor

310

by sulfated glycosaminoglycans, Cell 7 (1976) 85-95.

645. M.S. Brown and J.L. Goldstein, Analysis of a mutant stain of human fibroblasts with a defect in the internalization of receptor-bound low density lipoprotein, Cell 9 (1976) 663-674.

646. E. Vasile, M. Simionescu and N. Simionescu, Visualization of the binding, endocytosis and transcytosis of low-density lipoprotein in the arterial endothelium in situ, J. Cell Biol. 96 (1983) 1677-1689.

647. R. Hamanaka, K. Kohno, T. Seguchi, K. Okamura, A. Morimoto, M. Ono, J. Ogata and M. Kuwano, Induction of low density lipoprotein receptor and a transcription factor SP-1 by tumor necrosis factor in human microvascular endothelial cells, J. Biol. Chem. 267 (1992) 13160-13165.

648. M. Krieger and J. Herz, Structures and functions of multiligand lipoprotein receptors: Macrophage scavenger receptors and LDL receptor-related protein (LRP), Annu. Rev. Biochem. 63 (1994) 601-637.

649. B.W. Howell and J. Herz, The LDL receptor gene family: signaling functions during development, Curr. Opin. Neurobiol. 11 (2001) 74-81.

650. J. Heeren, T. Grewal, S. Jackle and U. Beisiegel, Recycling of apolipoprotein E and lipoprotein lipase through endosomal compartments in vivo, J. Biol. Chem. 276 (2001) 42333-42338.

651. G. Olivecrona and T. Olivecrona, Triglyceride lipases and athero-sclerosis, Curr. Opin. Lipidol. 6 (1995) 291-305.

652. R.W. Mahley and Z.S. Ji, Remnant lipoprotein metabolism: key pathways involving cell-surface heparan sulfate proteoglycans and apolipoprotein E, J. Lipid Res. 40 (1999) 1-16.

653. J.M. Felts, H. Itakura and R.T. Crane, The mechanism of assimilation of constituents of chylomicrons, very low density lipoproteins and remnants - a new theory, Biochem. Biophys. Res. Commun. 66 (1975) 1467-1475.

654. I.J. Goldberg, Lipoprotein lipase and lipolysis: central roles in lipoprotein metabolism and atherogenesis, J. Lipid Res. 37 (1996) 693-707.

655. A. Zambon, I. Schmidt, U. Beisiegel and J.D. Brunzell, Dimeric lipoprotein lipase is bound to triglyceride-rich plasma lipoproteins, J. Lipid Res. 37 (1996) 2394-2404.

656. J. Heeren and U. Beisiegel, Intracellular metabolism of trigly-ceride-rich lipoproteins, Curr. Opin. Lipidol. 12 (2001) 255-260.

657.	U. Beiseigel, W. Weber, G. Ihrke, J. Herz and K.K. Stanley, The LDL-receptor-related protein, LRP, is an apolipoprotein E-binding protein, Nature 341 (1989) 162-164.

658.	U. Beiseigel and J. Heeren, Lipoprotein lipase targeting of lipoproteins to receptors, Proc. Nutr. Soc. 56 (1997) 731-737.

659.	R.C. Kowal, J. Herz, J.L. Goldstein, V. Esser and M.S. Brown, Low density lipoprotein receptor related protein mediates uptake of cholesteryl esters derived from apoprotein E-enriched lipoproteins, Proc. Natl. Acad. Sci. USA 86 (1989) 5810-5814.

660.	W.A. Bradley and S.H. Gianturco, Apo E is necessary and sufficient for the binding of large triglyceride-rich lipoproteins to the LDL receptor; apo B is unnecessary, J. Lipid Res. 27 (1986) 40-48.

661.	A. Rohlmann, M. Gotthardt, R.E. Hammer and J. Herz, Inducible inactivation of hepatic LRP gene by Cre-mediated recombination confirms role of LRP in clearance of chylomicron remnants, J. Clin. Invest. 101 (1998) 689-695.

662.	T.E. Willnow, Z. Sheng, S. Ishibashi and J. Herz, Inhibition of hepatic chylomicron remnant uptake by gene transfer of a receptor antagonist, Science 264 (1994) 1471-1474.

663.	T.E. Willnow, A. Nykjaer and J. Herz, Lipoprotein receptors: new roles for ancient proteins, Nat. Cell Biol. 1 (1999) E157-E162.

664.	B.C. Mortimer, D.J. Beveridge, I.J. Martins and T.G. Redgrave, Intracellular localization and metabolism of chylomicron remnants in the livers of low density lipoprotein receptor-deficient mice and apo E-deficient mice, J. Biol. Chem. 270 (1995) 28767-28776.

665.	A.D. Cooper, Hepatic clearance of plasma chylomicron remnants, Semin. Liver Dis. 12 (1992) 386-396.

666.	D.A. Chappell and J.D. Medh, Receptor-mediated mechanisms of lipoprotein remnant catabolism, Prog. Lipid Res. 37 (1998) 393-422.

667.	M. Yamamoto, K. Ikeda, K. Ohshima, H. Tsugu, H. Kimura and M. Tomonaga, Increased expression of low density lipoprotein receptor-related protein/α2-macroglobulin receptor in human malignant astrocytomas, Cancer Res. 57 (1997) 2799-2805.

668.	G. Bu, E.A. Maksymovitch, H. Geuze and A.L. Schwartz, Sub-cellular localization and endocytic function of low density lipoprotein receptor-related protein in human glioblastoma cells,

312

J. Biol. Chem. 269 (1994) 29874-29882.
669.   L. Maletinska, E.A. Blakely, K.A. Bjornstad, D.F. Deen, L.J. Knoff and T.M. Forte, Human glioblastoma cell lines: levels of low-density lipoprotein receptor and low-density lipoprotein receptor-related protein, Cancer Res. 60 (2000) 2300-2303.
670.   Y. Li, N. Wood, P. Grimsley, D. Yellowlees and P.K. Donnelly, In vitro invasiveness of human breast cancer cells is promoted by low density lipoprotein receptor-related protein, Invasion Metastasis 18 (1998) 240-251.
671.   U.P. Steinbrecher, Receptors for oxidized low density lipoprotein, Biochim. Biophys. Acta 1436 (1999) 279-298.
672.   D. Rhainds, L. Falstrault, C. Tremblay and L. Brissette, Uptake and fate of class B scavenger receptor ligands in HepG2 cells, Eur. J. Biochem. 261 (1999) 227-235.
673.   S. Miki, A. Matsumoto, Y. Nakamura, H. Itakura, T. Kodama, M. Yamamoto and Y. Miki, Expression of scavenger receptors on renal cell carcinoma cells in vitro, Biochem. Biophys. Res. Commun. 189 (1992) 1323-1328.
674.   A. Mukhopadhyay, B. Mukhopadhyay, R.K. Srivastava and S.K. Basu, Scavenger-receptor mediated delivery of daunomycin elicits selective toxicity towards neoplastic cells of macrophage lineage, Biochem. J. 284 (1992) 237-241.
675.   B. Mukhopadhyay, A. Mukhopadhyay and S.K. Basu, Enhancement of tumouricidal activity of daunomycin by receptor-mediated delivery: in vivo studies, Biochem. Pharmacol. 46 (1993) 919-924.
676.   S. Basu, B. Mukhopadhyay, S.K. Basu and A. Mukhopadhyay, Enhanced intracellular delivery of doxorubicin by scavenger receptor-mediated endocytosis for preferential killing of histiocytic lymphoma cells in culture, FEBS Lett 342 (1994) 249-254.
677.   M.D. Bell, R. Lopez-Gonzalez, L. Lawson, D. Hughes, I. Fraser, S. Gordon and V.H. Perry, Upregulation of the macrophage sacvenger receptor in response to different forms of injury in the CNS, J. Neurocytol. 23 (1994) 605-613.
678.   M. Lucarelli, M. Gennarelli, P. Cardelli, G. Novelli, S. Scarpa, B. Dallapiccola and R. Strom, Expression of receptors for native and chemically modified low-density lipoproteins in brain microvessels, FEBS Lett. 401 (1997) 53-58.
679.   A. Bignami and D. Dahl, The astroglial response to stabbing: immunofluorescence studies with antibodies to astrocyte-specific

protein (GFAP) in mammalian and submammalian vertebrates, Neuropathol. Appl. Neurobiol. 2 (1976) 99-110.

680. A.J. Mathewson and M. Berry, Observation on the astrocyte response to a cerebral stab wound in adult rats, Brain Res. 327 (1985) 61-69.

681. T. Miyake, M. Okada and T. Kitamura, Reactive proliferation of astrocytes studied by immunohistochemistry for proliferating cell nuclear antigen, Brain Res. 590 (1992) 300-302.

682. M. Nieto-Sampedro, R.P. Saneto, J. de Vellis and C.W. Cotman, The control of glial populations in brain: changes in astrocyte mitogenic and morphologic factors in response to injury, Brain Res. 343 (1985) 320-328.

683. D. Schiffer, M.T. Giordana, A. Migheli, G. Giaccone, S. Pezzotta and A. Mauro, Glial fibrillary acidic protein and vimentin in the experimental glial reaction of the brain, Brain Res. 374 (1986) 110-118.

684. F.J. Liuzzi and R. Lasek, Astrocytes block axonal regeneration in mammals by activating the physiological stop pathway, Science 237 (1987) 642-645.

685. M. Gimenezy Ribotta, N. Rajaofetra, C. Morin-Richaud, G. Alonso, A.D. Bochelen, F. Sandillon, A. Legrand, M. Mersel and A. Privat, Oxysterol (7β-hydroxycholesteryl-3-oleate) promotes serotonergic reinnervation in the lesioned rat spinal cord by reducing glial reaction, J. Neurosci. Res. 41 (1995) 79-95.

686. L. Meda, M.A. Cassatella, G.I. Szendrel, L. Otvos, P. Baron, M. Villalba, D. Ferrari and F. Rossi, Activation of microglial cells by β-amyloid protein and interferon-γ, Nature 374 (1995) 647-650.

687. J. El Khoury, S.E. Hickman, C.A. Thomas, L. Cao, S.C. Silverstein and J.D. Loike, Scavenger receptor-mediated adhesion of microglia to β-amyloid fibrils, Nature 382 (1996) 716-719.

688. K.K. Kopec and R.T. Carroll, Alzheimer's beta-amyloid peptide 1-42 induces a phagocytic response in murine microglia, J. Neurochem. 71 (1998) 2123-2131.

689. D.M. Paresce, R.N. Ghosh and F.R. Maxfield, Microglial cells internalize aggregates of the Alzheimer's disease amyloid β-protein via a scavenger receptor, Neuron 17 (1996) 553-565.

690. R.H. Christie, M. Freeman and B.T. Hyman, Expression of the macrophage scavenger receptor, a multifunctional lipoprotein receptor, in microglia associated with senile plaques in

314

Alzheimer's disease, Am. J. Pathol. 148 (1996) 399-403.

691.   J. El Khoury, S.E. Hickman, J.D. Thomas, J.D. Loike and S.C. Silverstein, Microglia, scavenger receptors, and the pathogenesis of Alzheimer's disease, Neurobiol. Aging 19 (1998) S81-S84.

692.   R.N. Kalaria, Microglia and Alzheimer's disease, Curr. Opin. Hematol. 6 (1999) 15-24.

693.   D.M. Watterson, S. Mirzoeva, L. Guo, A. Whyte, J.J. Bourguignon, M. Hibert, J. Haiech and L.J. Van Eldik, Ligand modulation of glial activation: cell permeable, small molecule inhibitors of serine-threonine protein kinases can block induction of interleukin 1 beta and nitric oxide synthase II, Neurochem. Int. 39 (2001) 459-468.

694.   L. Guo, A. Sawkar, M. Zasadzki, D.M. Watterson and L.J. Van Eldik, Similar activation of glial cultures from different rat brain regions by neuroinflammatory stimuli and downregulation of the activation by a new class of small molecule ligands, Neurobiol. Aging 22 (2001) 975-981.

695.   A.G. Lam, T. Koppal, K.T. Akama, L. Guo, J.M. Craft, B. Samy, J.P. Schavocky, D.M. Watterson and L.J. Van Eldik, Mechanism of glial activation by S100B: involvement of the transcription factor NFkappaB, Neurobiol. Aging 22 (2001) 765-772.

696.   S. Mirzoeva, A. Zawkar, M. Zasadzki, L. Guo, A.V. Valentza, V. Dunlap, J.J. Bourguignon, H. Ramstrom, J. Haiech, J.J. Van Eldik and D.M. Watterson, Discovery of a 3-amino-6-phenyl-pyridazine derivative as a new synthetic antineuroinflammatory compound, J. Med. Chem. 45 (2002) 563-566.

697.   S.K. Malhotra, T.K. Shnitka and J. Elbrink, Reactive astrocytes - a review, Cytobios 61 (1990) 133-160.

698.   J. El Khoury, S.E. Hickman, C.A. Thomas, H. Suzuki, T. Kodama and S.C. Silverstein, Class A scavenger receptors mediate adhesion of microglia to a neurotoxic fragment of prion protein, Conference Proceedings of Biomedicine '98, Washington DC, 1998.

699.   Y. Yamada, T. Doi, T. Hamakubo and T. Kodama, Scavenger receptor family proteins: roles for atherosclerosis, host defense and disorders of the central nervous system, Cell Mol. Life Sci. 54 (1998) 628-640.

700.   H. Shirai, T. Murakami, Y. Yamada, T. Doi, T. Hamakubo and T. Kodama, Structure and function of type I and II macrophage scavenger receptors, Mech. Ageing Dev. 111 (1999) 107-121.

701. P.J. Gough, D.R. Greaves, H. Suzuki, T. Hakkinen, M.O. Hiltunen, M. Turunen, S.Y. Herttuala, T. Kodama and S. Gordon, Analysis of macrophage scavenger receptor (SR-A) expression in human aortic atherosclerotic lesions, Arterioscler. Thromb. Vasc. Biol. 19 (1999) 461-471.

702. M.F. Linton and and S. Fazio, Class A scavenger receptors, macrophages, and atherosclerosis, Curr. Opin. Lipidol. 12 (2001) 489-495.

703. K. Hayashida, N. Kume, M. Minami and T. Kita, Lectin-like oxidized LDL receptor-1 (LOX-1) supports adhesion of mononuclear leukocytes and a monocyte-like cell line THP-1 cells under static and flow conditions, FEBS Lett. 511 (2002) 133-138.

704. G. Draude and R.L. Lorenz, TGF-β1 downregulates CD36 and scavenger receptor A but upregulates LOX-1 in human macrophages, AJP - Heart Circ. Physiol. 278 (2000) H1042-H1048.

705. A.C. Nicholson, M. Febbraio, J. Han, R.L. Silverstein and D.P. Hajjar, CD36 in atherosclerosis: the role of a class B macrophage scavenger receptor, Ann. NY Acad. Sci. 902 (2000) 128-131.

706. K. Hirano, S. Yamashita, Y. Nakagawa, T. Ohya, F. Matsuura, T. Tsukamoto, Y. Okamoto, A. Matsuyama, K. Matsumoto, J. Miyagawa and Y. Matsuzawa, Expression of human scavenger receptor class B type I in cultured human monocyte-derived macrophages and atherosclerotic lesions, Circ. Res. 85 (1999) 108-116.

707. G. Chinetti, F.G. Gbaguidi, S. Griglio, Z. Mallat, M. Antonucci, P Poulain, J. Chapman, J.C. Fruchart, A. Tedgui, J. Najib-Fruchart and B. Staels, CLA-1/SR-BI is expressed in atherosclerotic lesion macrophages and regulated by activators of peroxisome proliferator-activated receptors, Circulation 101 (2000) 2411-2417.

708. Y. Nakagawa-Toyama, S. Yamashita, J. Miyagawa, M. Nishida, S. Nozaki, H. Nagaretani, N. Sakai, H. Hiraoka, K. Yamamori, T. Yamane, K. Hirano and Y. Matsuzawa, Localization of CD36 and scavenger receptor class A in human coronary arteries - a possible difference in the contribution of both receptors to plaque formation, Atherosclerosis 156 (2001) 297-305.

709. A.C. Sposito, R.D. Santos, W. Hueb, L.I. Ventura, C.C.G. Vinagre, J.A.F. Ramires and R.C. Maranhao, LDL concentration is correlated with the removal from the plasma of a chylomicron-like emulsion in subjects with coronary artery disease,

316

Atherosclerosis 161 (2002) 447-453.

710.   M.A. Longino, D.A. Bakan, J.P. Weichert and R.E. Counsell, Formulation of polyiodinated triglyceride analogues in a chylomicron remnant-like liver-selective delivery vehicle, Pharm. Res. 13 (1996) 875-879.

711.   D.A. Bakan, M.A. Longino, J.P. Weichert and R.E. Counsell, Physicochemical characterization of a synthetic lipid emulsion for hepatocyte-selective delivery of lipophilic compounds: application to polyiodinated triglycerides as contrast agents for computed tomography, J. Pharm. Sci. 85 (1996) 908-914.

712.   T. Hara, Y. Tan and L. Huang, In vivo gene delivery to the liver using reconstituted chylomicron remnants as a novel nonviral vector, Proc. Natl. Acad. Sci. USA 94 (1997) 14547-14552.

713.   P.C.N. Rensen, N. Herijgers, M.H. Netscher, S.C.J. Meskers, M. van Eck and T.J.C. van Berkel, Particle size determines the specificity of apolipoprotein E-containing triglyceride-rich emulsions for the LDL-receptor versus hepatic remnant receptor in vivo, J. Lipid Res. 38 (1997) 1070-1084.

714.   W.A. Bradley and S.H. Gianturco, Apo E is necessary and sufficient for the binding of large triglyceride-rich lipoproteins to the LDL receptor; apo B is unnecessary, J. Lipid Res. 27 (1986) 40-48.

715.   E. Wisse, R.B. de Zanger, K. Charels, P. van der Smissen and R.S. McCuskey, The liver sieve: considerations concerning the structure and function of endothelial fenestrae, the sinusoidal wall and the space of Disse, Hepatology 5 (1985) 683-692.

716.   F. Braet, W.H.J. Kalle, R.B. de Zanger, B.G. de Grooth, A.K. Raap, H.J. Tanke and E. Wisse, Comparative atomic force and scanning electron microscopy: an investigation on fenestrated endothelial cells in vitro, J. Microscopy 181 (1996) 10-17.

717.   D.C. Grant, Retention of lipid-coated microbubbles by membrane filters, Report issued 7-19-02 by CT Associates Inc., 2002, 15 pp. (limited distribution report).

SUBJECT INDEX

320

322

T